CONTEMPORARY RURAL SYSTEMS IN TRANSITION

Volume 2

Economy and Society

CONTEMPORARY RURAL SYSTEMS IN TRANSITION

VOLUME 2
Economy and Society

Edited by

I.R. Bowler, C.R. Bryant and M.D. Nellis

C·A·B International

C·A·B International   Tel: Wallingford (0491) 32111
Wallingford   Telex: 847964 (COMAGG G)
Oxon OX10 8DE   Telecom Gold/Dialcom: 84: CAU001
UK   Fax: (0491) 33508

A catalogue record for this book is available from the British Library

ISBN 0 85198 811 3 (Volume 1)
ISBN 0 85198 812 1 (Volume 2)

ISBN 0 85198 813 X (Two-volume set)

Phototypeset by Intype, London
Printed and bound in the UK by Redwood Press Ltd, Melksham

CONTENTS

V Development Strategies for Rural Communities

NOTES ON CONTRIBUTORS

Robert Annis is a crosscultural psychologist who is currently Executive Director of WESTARC Group Inc., the applied research arm of Brandon University, Brandon, Manitoba, a position he has held since 1987. Prior to that time he was an Associate Professor in the Department of Native Studies at Brandon University, and is currently series editor of *Abstracts of Native Studies*. His academic interests lie within cross-cultural psychology, as well as in the serious issues that are threatening the contemporary rural communities of the Prairie Provinces.

Clark Archer is Associate Professor of Geography at the University of Nebraska-Lincoln. He received a BA in Political Science (1965) and an MA in Geography (1968) from Indiana University; he also received a PhD in Geography (1974) from the University of Iowa. His teaching and research interests involve political geography, population geography, cartography, and spatial analysis. He is coauthor, with P. Taylor, of *Section and Party* (1981, Wiley, New York) and, with F. Shelley, of *American Electoral Mosaics* (1986, Association of American Geographers, Washington DC).

Ray Bollman is a research economist with Statistics Canada. His research has been focused on farm and off-farm family income, entry, exit and structural change in agriculture and trends and characteristics of rural populations. He has served as President of the Canadian Association of Rural Studies, as President of the Canadian Agricultural Economics and Farm Management Society and as co-organizer of the Agricultural and Rural Restructuring Group.

Ian Bowler is a Senior Lecturer in the Department of Geography at the University of Leicester, having previously worked at the University of Strathclyde. He holds BA and PhD degrees from the University of

Liverpool and has a specialist research interest in state intervention in agriculture. His books include *Agriculture under the Common Agricultural Policy* (1986, Manchester University Press, Manchester), and *The Geography of Agriculture in Developed Market Economies* (Editor) (1993, Longman, London). He is a former Chair of the Rural Geography Study Group of the Institute of British Geographers and a member of the IGU Commission on 'Changing Rural Systems'.

Christopher Bryant has been a full professor in the Département de Géographie at the Université de Montréal since 1990, after being with the University of Waterloo in southern Ontario since 1970; he was also Director of the Economic Development Programme at Waterloo between 1984 and 1990. He holds a PhD from the London School of Economics and Political Science. His primary research interests are urban fringe agriculture and local and community economic development. Together with L.H. Russwurm and A.G. McLellan, he published *The City's Countryside* in 1982 (Longman, London) and *Agriculture in the City's Countryside* in 1992, co-authored with T.R.R. Johnston (Belhaven Press, London).

Michael Bunce received his BA (1966) and his PhD (1970) from the University of Sheffield, UK. He is now Associate Professor in Geography at Scarborough College, University of Toronto. His research interests include rural and agricultural land use, land use change in the urban fringe, rural planning and attitudes to the countryside. Major publications include *Rural Settlement in an Urban World* (1982, Croom Helm, London) and *The Pressures of Change in Rural Canada* (1984, Geographical Monograph 14, York University, co-edited with M.J. Troughton). From 1985 to 1991, he was Chair of the Rural and Urban Fringe Study Group, Canadian Association of Geographers.

Richard Butler was educated at Nottingham and Glasgow Universities and has taught at the University of Western Ontario since 1967. He is currently President of the Canadian Association for Leisure Studies, a past Secretary of the International Academy for the Study of Tourism and past President of the Ontario Research Council on Leisure. His principal research interests are in tourism and recreation with a particular focus on the process and effects of development of destination areas, using examples from North America, Western Europe and the Caribbean, and with the effects of development in general on peripheral and insular locations. He has published widely on tourism and is best known for his theoretical and conceptual contributions towards a cycle of evolution of destination areas. His present research deals with the

problems of evaluating and controlling the impacts of leisure on destination communities and environments.

Gordon Clark MA, PhD (Edinburgh) is Senior Lecturer in Geography at
Lancaster University and currently Chair of the Rural Geography Study
Group of the Institute of British Geographers. His research interests
are in the area of rural geography including the restructuring of agriculture, rural planning, innovation diffusion and housing issues in the
countryside. Much of this research has focused on Scotland and northwest England. He has longstanding research links with Dutch rural
geographers and is the author of *Housing and Planning in the Countryside* (1982, Wiley, Chichester).

Paul Cloke is Reader in Geography at the University of Bristol. He is
editor of the *Journal of Rural Studies* and has been author of several
books on rural geography, including *The Rural State? Limits to Planning
in Rural Society* (with Jo Little, 1990, Clarendon Press, Oxford) and
Policy and Change in Thatcher's Britain (1992, Pergamon, Oxford). He
is currently engaged on a programme of research into changing rural
life-styles in England and Wales.

Nigel Curry BA, MSc, PhD, MRTPI, MILAM is Professor of Countryside
Planning at the Cheltenham and Gloucester College of Higher Education, England, and is Visiting Professor of Environmental Planning
at the Queen's University of Belfast, Northern Ireland. His research
interests include countryside recreation, the structure of the rural economy, the planning process in the countryside and village appraisals. He
has edited three general texts on the British countryside: *The Changing
Countryside* (1986, Open University, London) *A Future for Our
Countryside* (1988, Blackwell, Oxford) and *A People's Charter?* (1990,
HMSO, London) and has authored a text *Countryside Recreation:
Provision, Participation, Preference, Plans and Policies* due for publication in 1992.

Floyd Dykeman is Director of the Rural and Small Town Research and
Studies Programme, Mount Allison University, New Brunswick. He has
worked as a planner in New Brunswick and British Columbia and has
provided consulting services in a number of provinces for both the
public and private sectors. His research and practising interests include
integrated rural planning systems, rural economic and small business
development, strategic planning, innovation in planning and tourism
planning. His most recent publications include edited works entitled
Integrated Rural Planning and Development (1988, Mount Allison

University, New Brunswick) and *Entrepreneurial and Sustainable Rural Communities* (1990, Mount Allison University, New Brunswick).

John Everitt is Professor and Chairman of the Department of Geography at Brandon University, Brandon, Manitoba. His academic interests lie within the general area of human geography, and recently have centred upon applied research within the Prairie Provinces. He has recently looked at regional planning, various issues with regard to native peoples, the life satisfaction and migration of the elderly, and small community analysis. He has published (with others) four volumes of an *Inventory of Archival Material* (of Manitoba), Brandon University Press, two volumes of *Prairie Conference Proceedings*, Brandon University Press, and in 1988 was a co-author of *Brandon: Geographical Perspectives on the Wheat City*, CPRC, Regina. He has worked as a Senior Consultant for WESTARC Group Inc., the applied research arm of Brandon University, since 1986.

Paul Frederic is Professor of Geography at the University of Maine-Farmington. He has conducted research in the areas of rural land-use policy and United States–Canadian relations. His publications include several monographs and book chapters on land use and natural resource management. Frederic served as Executive Director of the Maine Land Use Regulation Commission, 1987–89, and in 1990 was appointed to the endowed Libra Professor Chair at the University of Maine-Farmington. He is also past President of the New England-St Lawrence Valley Division of the Association of American Geographers.

Anthony Fuller is a Professor in the University School of Rural Development and Planning, University of Guelph. He holds a PhD from the University of Hull, UK, where he worked on the transformation of farming in the Appennines in Northern Italy. At the University of Guelph, he was Director of the Rural Development Outreach Project which extended the services of the University to rural communities in Ontario. He is Research Advisor to the Arkleton Trust for a 5-year research programme on 'Structural Change in Agriculture and Pluriactivity Among Farm Families', 1986–1992, involving 24 study areas in 12 West European countries. He is also currently co-leader of the Canadian Agricultural and Rural Restructuring Group.

Owen Furuseth is a Professor in the Department of Geography and Earth Sciences at the University of North Carolina at Charlotte. He is a graduate of East Carolina University and Oregon State University. Furuseth's research interests include rural land use planning and policy and social relations in rural community decision making. He is the co-author of two books on environmental impact assessment and rural land use policy. He is a former regional editor of the *International Yearbook of Rural Planning* and is currently an editor of *Progress in Rural Policy and Planning*. In 1990 he was awarded the *John Fraser Hart Award for Research Excellence* by the Contemporary Agriculture and Rural Land Use Speciality Group of the Association of American Geographers.

Sue Glyptis is Professor of Recreation Management at Loughborough University. A geographer by training, her main research interests are in countryside recreation, leisure life-styles, and recreational disadvantage. Her recent publications include *Countryside Recreation* (1991, Longman, London), *Developing Sport and Leisure: Good Practice in Urban Regeneration* (1989, HMSO, London) and *Leisure and Unemployment* (1989, Open University Press, Milton Keynes). Prior to working at Loughborough she was Research Officer to The Sports Council (London), involved particularly in research on countryside sport, sports participation and facility management.

Sarah Harper is a Lecturer in Geography at Royal Holloway, University of London. She has worked in crosscultural gerontology since 1985, holding visiting positions in Gerontology Centres at the Universities of Utah and Southern California (USA), and La Trobe University, Melbourne (Australia). She is currently working on a major survey *Ageing Societies* – an international examination of the societal implications of ageing. She is the International Officer for the British Society of Gerontology, sits on the National Councils of the Institute of British Geography and British Society of Gerontology, and represents the UK on the Social and Behavioural Committee of the International Association of Ageing (Europe). She is a member of the Rural Geography Study Group (IBG) and a former committee member.

Glenda Laws (PhD, McMaster) is Assistant Professor at the Pennsylvania State University, Department of Geography. Her research interests focus on North American service-dependent populations and their demands for housing and support services. She is currently working on a project concerning the social construction of old age.

Fiona Mackenzie is Associate Professor with the Department of Geography at Carleton University. Her research interests include local restructuring among agricultural communities in Ontario and Québec near Ottawa; she also has research interests in Ghana relating to gender and land rights in a context of legal plurality, and in Kenya relating to the political economy of the environment, gender and resistance. Recent publications include *Development from Within: Survival in Rural Africa*

(1992, Routledge, London, co-edited with D.R.F. Taylor) and 'Political economy of the environment, gender and resistance' in the *Canadian Journal of African Studies* (1991).

Stephen Martin is a Lecturer in Public Sector Management at Aston Business School in Birmingham, UK. His main research interests are in the assessment of the impacts of public policy on rural areas and in the evaluation of the effects of urban regeneration programmes on inner cities. He has written widely on the role and activities of the Rural Development Commission and has recently undertaken major reviews of the Commission's Rural Development Programme Process (published in 1990) and of its Social and Community Development Programme (published in 1991). He is co-author of *Small Firms Policy and the Rural Environment* (1993, Routledge, Kegan and Paul, London) and is currently investigating the implications for the local economic development strategies of urban and rural local authorities in the UK of the reform of the EC Commission's Structural Funds.

Robert Mason is Assistant Professor of Geography at Temple University in Philadelphia, Pennsylvania, US. He is the author of *Contested Lands: Conflict and Compromise in New Jersey's Pine Barrens* (1992, Temple University Press, Philadelphia) and co-author of the *Atlas of United States Environmental Issues* (1990, Macmillan, New York). He continues to conduct research on the roles of parks, preserves and land trusts in sustaining rural landscapes, and he is also interested in community response to threats posed by routine and accidental releases of toxic chemicals.

Duane Nellis is currently Professor and Head of Geography, Presidential Lecturer, and Director of the Institute of Social and Behavioral Research at Kansas State University. Born in Spokane, Washington, Nellis received his BSc degree in geography from Montana State University in 1976, MSc and PhD from Oregon State University in 1977 and 1980 respectively. He has authored numerous articles on United States and African rural resource systems, and the application of remote sensing and geographic information systems to rural resource analysis. He has received numerous awards, including the Institute of British Geographers Young Research Worker Award, the Association of American Geographers Contemporary Agriculture and Rural Land Use Speciality Group *John Fraser Hart Award for Research Excellence*. He is also past Chair of the Contemporary Agriculture and Rural Land Use Speciality Group of the AAG.

Lizbeth Pyle was awarded her PhD in Geography at the University of

Minnesota and has been an Assistant Professor of Geography at West Virginia University for the past 6 years. Her broad teaching and research interests include issues in rural resource management, such as economic valuation of recreation resources and rural land markets. She has published her research in the *Geographical Review*, the *Professional Geographer, Urban Geography*, and the *Southeastern Geographer*.

Guy Robinson is Senior Lecturer in Human Geography at the University of Edinburgh. His research interests include: rural development in the Developed World, managing urban sprawl in Canada and Australasia, and the commercialization of agriculture in Central America. His recent books are: *Conflict and Change in the Countryside* (1990, Belhaven Press, London); *Agricultural Change: Geographical Studies of British Agriculture* (1988, North British Publishing, Edinburgh); (editor) *A Social Geography of Canada* (revised edition, 1991, North British Publishing, Edinburgh).

Nigel Walford has been a Senior Lecturer in the School of Geography, Kingston Polytechnic, since 1989. He was previously Chief Research Officer in the ESRC Data Archive at the University of Essex where he latterly worked on the creation of the Rural Areas Database. He has contributed chapters to a number of books including, most recently, *People in the Countryside* (eds A. Champion and C. Watkins, 1991, Paul Chapman, London); *Rural France and Great Britain* (ed. P. Brunet, 1991, University of Caen, Caen); and *Restructuring the Countryside: Environmental Policy in Practice* (ed. A. Gilg, 1992, Avebury, Aldershot, UK).

Gerald Walker is Associate Professor in the Department of Geography, York University, Canada. His principal research interests deal with the transformation of rural life in the countryside, particularly in the context of Toronto's countryside. His research has emphasized the social recomposition of the countryside and the importance of understanding social networks and relationships. He has also developed an active research programme in the area of land-use change. Among his recent publications are *An Invaded Countryside* (Geographical Monograph 17, York University) and *The Farmers' Class* (Geography Discussion Paper 34, York University).

Jackie Wolfe is Associate Professor in the University School of Rural Planning and Development at the University of Guelph. Her major research interests are in local level development and community planning and decision making by the First Peoples in Canada and Australia, with a focus on traditional and conventional planning processes and models; local government and self-government; indigenous and western ecological knowledge and land and resource management systems; micro-enterprises; community services delivery; and women's roles in community decision making. Publications include *That Community Government Mob*: *Local Government in Small Northern Territory Communities* (1989, Australian National University North Australia Research Unit, Darwin).

PREFACE

In August 1991, an international conference was convened in the United Kingdom by Dr Ian Bowler; the conference brought together rural geographers from Canada (coordinator: Michael Bunce), the United Kingdom and the United States (coordinator: Duane Nellis). The focus of the conference was the rapid and significant transition of rural areas in developed countries over the previous decade, a transition sometimes referred to as the 'restructuring' of rural economy and society. The conference reviewed a wide range of issues as regards contemporary changes in rural areas, and adopted the descriptive term 'rural system' to encompass and integrate the totality of rural economy and society.

The two volumes of this book are based on the papers presented at the conference. Volume 1 contains papers on agriculture and the environment: (i) developing organizational structures in the food supply system; (ii) changing farm business structures, (iii) the environmental impact of modern agriculture; (iv) emerging agricultural policy issues; and (v) the growing concern with 'sustainable agriculture'. In Volume 2, the papers move on to consider aspects of the wider rural economy and society: (i) the emergence of new socioeconomic issues, (ii) changes in the structure of rural society, (iii) trends in countryside recreation and tourism, (iv) the changing employment structure, and (v) development strategies for rural communities.

Each chapter in the book discusses broad processes and structural changes that are common to all rural systems in developed countries; however, different geographical contexts are used to illustrate the uneven development of those systems under the contemporary transition of rural areas. Contributors have drawn upon their varying research experiences in Canada, the UK and the US: they review the relevant literature for their national contexts and illustrate their arguments from original research. The resulting book, in its two volumes, covers a wide range of issues in the contemporary transition of rural systems and provides a contrasting set of international perspectives.

ACKNOWLEDGEMENT

The editors wish to thank the Economic and Social Research Council (UK) for financial support of the conference, the papers from which form the basis of this book.

ABBREVIATIONS

AAG	Association of American Geographers
ACOA	Atlantic Canada Opportunities Agency
AFA	Alternative Funding Arrangements
AFDC	Aid to Families with Dependent Children
ALURE	Alternative Land Use and the Rural Economy
AONB	Area of Outstanding Natural Beauty
BFAP	Beginning Farmer Assistance Programme
BTA	British Tourist Authority
CA	Census Agglomeration
CAP	Common Agricultural Policy
CCC	Commodity Credit Corporation
CCPR	Central Council for Physical Recreation
CMA	Census Metropolitan Area
C$	Canadian dollar
CUSTA	Canada–United States Trade Agreement
DIAND	Department of Indian Affairs and Northern Development
DoE	Department of the Environment (UK)
DoEC	Department of Environmental Conservation (New York State)
EC	European Community
ECU	European Currency Unit
ESU	European Size Unit
FCC	Farm Credit Corporation
FTE	Full Time Equivalent
GATT	General Agreement on Tariffs and Trade
GDP	Gross Domestic Product
GNP	Gross National Product
ha	hectares
HC	Housing Corporation
HIP	Housing Investment Programmes

HMSO	Her Majesty's Stationery Office
INAC	Indian and Northern Affairs Canada
JBNQA	James Bay and Northern Quebec Agreement
LDSPB	Lake District Special Planning Board
LWCF	Land and Water Conservation Fund
MA	Metropolitan Area
MAFF	Ministry of Agriculture Fisheries and Food
MSA	Metropolitan Statistical Area
NNR	National Nature Reserve
NACRT	National Agricultural Centre Rural Trust
NAN	Nishnawbe-Aski Nation
NIMBY	Not in my back-yard
NPAC	National Parks and Access to the Countryside Act (UK)
NWT	Northwest Territories
OECD	Organisation for Economic Cooperation and Development
OPCS	Office of Population Census and Surveys
PEI	Prince Edward Island
PLI	Persistently low income
RDC	Rural Development Commission
RSPB	Royal Society for the Protection of Birds
SGM	Standard Gross Margin
SSSI	Site of Special Scientific Interest
TVA	Tennessee Valley Authority
UK	United Kingdom
US(A)	United States (of America)
US$	United States dollar
WSA	Women for the Survival of Agriculture

INTRODUCTION

Rural areas have been undergoing rapid change as their economies and societies restructure, partially at least in response to changes in the broader socioeconomic system. Structures and internal and external relationships have been modified in relation to powerful forces that transcend all scales of analysis. In Volume 1, the accent was placed on changing agricultural structures and relationships, and a significant thread dealt with the links between agriculture and environment. In Volume 2, the emphasis shifts from farming, which remains the major activity and land use in many rural areas, to a consideration of certain aspects of the broader rural economy and society.

Two recurrent themes run through most discussions of rural change: uneven development and the consequences of and responses to uneven development. Uneven development, in terms of differing levels of socio-economic health, arises from processes of capital accumulation and geographic and non-geographic concentrations of power and wealth, all in the context of limited financial resources. Uneven development is fuelled by flows of capital and labour, and is reinforced by uneven patterns of competitiveness and flows of goods and services. It is manifested in geographic and non-geographic concentrations of wealth and power, and their opposites, poverty, disenfranchisement and lack of control. Patterns and processes of uneven development have always existed, but in the last quarter of the 20th century a critical question is posed for Western countries: that is whether the patterns of uneven development are strengthening and becoming more accentuated. In the context of rural areas, which have long been associated with economic stagnation in the more peripheral parts of the economic system, are these areas becoming increasingly vulnerable? Are their economic and social structures more and more marginalized? Furthermore, what sorts of responses and strategies have been followed to try and modify the internal and external relationships of rural areas? And what sorts of impacts have they had?

1

Although the chapters in the two volumes of this book do not address the whole range of issues, they do touch on most of them. This introductory chapter develops a framework with which to view the contemporary shifts in the rural economies and societies of the Western world and which serves as a backdrop to the contributions in this Volume.

The dynamics of rural change and uneven development are considered first, emphasizing the ties between rural areas and the other parts of our social and economic systems. This theme is also reiterated in a discussion of the heterogeneous nature of rural areas, and the different forces of change that have influenced their transformation. With this as a backdrop, consideration is given to the results of these transformations and some selected responses to them, drawing upon the various contributions to Volume 2.

The Dynamics of Rural Change and Uneven Development

Underlying uneven development processes are changes in activities – economic, social, political – that result from modification of socioeconomic systems of production and of the systems of exchange within which all activities function. Different socioeconomic systems of production are characterized by different technical, social, financial, economic and behavioural structures and patterns of interaction; different systems dominate at different times: for example, at the coarsest level of classification, we can identify subsistence production, artisanal production, capitalistic production and more collectivist modes of organization of production (see Malassis, 1958).

Systems change and, over longer periods of time, are replaced. When the seeds of a newly emerging system are concentrated geographically, the scene is ripe for differential development of the dominant system over geographic space and through time. Hence, at least one of the types of areas where we might expect to see significant developments in capitalistic forms of rural activities is in close proximity to the urban-industrial complex.

Socioeconomic systems of production function through various systems of exchange or interaction (for example, market systems for produce, capital or labour), and these systems can function at a variety of different scales. As the macrosystems change, so do the systems of exchange that animate them, and vice versa. Resulting from this are changing relationships between rural areas (and their activities) and other parts of the economic and social system.

The Heterogeneity of Rural Areas

Many discussions of the nature of change in rural economies and societies oversimplify the situation by talking of 'rural' areas as if they were stand-alone entities and as if they were homogeneous. In relation to these two points, the following should be noted.

1. A 'rural' area is not necessarily a functioning socioeconomic or political unit. At the very least, it is important to appreciate the links that exist between many rural areas and the small towns embedded in them, frequently serving as service centres for rural market areas.
2. Furthermore, 'rural' areas can be differentiated on the basis of a number of dimensions, which are essential for understanding the results of and responses to the transformation of rural areas. Some of the more important differentiating characteristics of rural areas are: the degree of integration of rural areas with the urban–industrial complex; the nature of the economic base of the area (for example, the natural resource base, tourism, services and manufacturing; the social and demographic structure of the area and the community or communities present; and the nature of the political organization of the area and the aspirations and capacities of the communities in relation to their abilities for self-governance and management).

The principal reason for emphasizing the heterogeneity of rural areas is because such differences can be critical in understanding patterns of responses from rural areas and the appropriate strategies for dealing with 'rural' issues. Furthermore, not only is there heterogeneity among rural areas, there are also significant differences between groups of 'rural' people even within the same geographic community or locally. Once again, these differences are important in understanding issues, patterns of responses and appropriate strategies, both at the individual or micro level and at the community and state level.

The Forces of Change

Fundamental to the notion of systems of exchange introduced above is the idea that the interactions reflect the supply and demand for different types of interaction. In the context of economic systems of exchange, different types of interaction reflect the values placed on the goods and services involved in those interactions; the functioning of economic systems of exchange is invariably associated with different socioeconomic modes of production and can also have impacts on social and political systems of exchange. For instance, the operating of the product, labour and capital markets leads to uneven development geographically, giving rise to changes in social systems of exchange within rural communities where there is out-migration and the development of pockets of poverty.

At the most general level, interactions can be modified by changes in values, institutions and technologies. Changes in values undoubtedly constitute the most fundamental set of changes, being tied to changing demographic structures and migration patterns, as well as underlying the other two categories – institutional and technological changes. During the 1980s, it became fashionable to talk of the broad forces of change in society as 'mega-trends' (see Naisbitt, 1982), particularly in relation to the transformation of society from an industrial to an information and services, or postindustrial, society.

It is easy to think of these forces or trends as somehow removed from the individual, the firm or farm and the rural community, and therefore to think of these as requiring some adjustment, reaction or adaption on the part of the individuals, firms, farms and communities. However, and this is important from the rural area or community perspective, it should not be forgotten that such trends are composed of the decisions of many individuals and firms. Influences can be transmitted upwards through the systems of exchange as well as downwards, thus emphasizing the opportunity for some proactive behaviour on the part of the elemental decision-making units in the socioeconomic system. Of course, the opportunities for engaging in proactive behaviour vary between rural areas of communities and between individuals in those areas.

What have been the main forces of change leading to transitions in the social and economic system and how have these been translated into the restructuring of rural systems? It is useful to think of the various forces or trends associated with the development of postindustrial society as falling into three categories: (i) the emergence of 'new needs' in society; (ii) the development of new transportation and communications technology; and (iii) the development of new technology affecting production processes (Bryant, 1989). All of these forces have influenced the restructuring of rural systems, as well as many other facets of the organization of society and economy.

'New needs' are linked to a variety of value changes that have occurred and continue to occur in society. As disposable income has increased and the population has, on average, become more 'sophisticated', demands have increased for a greater range of services as well as for goods with a considerable value-added component to them. Values regarding life-styles have changed with, for instance, greater values being placed on the quality of the rural environment, outdoor recreation and living in a rural community. One outcome has been a resurgence of interest in many countries during the 1980s in the potential for tourist development to bring greater income security to many rural communities. Other value changes include the greater attention being paid to personal health, to the quality of food products and to the environment generally. All of these, potentially, have impacts on rural socioeconomic systems. Some of these changes have the potential to contribute to the rejuvenation of the economic systems of some rural areas, whereas in others they can create difficulties. One need only think of the developing market for paper produced by processes that are chlorine-free to realize the far-reaching impact this will have on the paper industry and the resource hinterlands involved, depending on the ability of the companies to modify or change their production technology.

It is also possible to include under new needs the pleas for deregulation and less direct government intervention in many countries. In North America, the Canada–US Free Trade Agreement (CUSTA) can perhaps be placed in this context, and it certainly has far-reaching implications for many activities in rural areas. Similarly, the latest round of negotiations concerning GATT hold significant implications for agricultural activities in many countries.

Changes in transportation and communications technology have been central to the development of postindustrial society. An important part of the changes in transportation technology is, of course, intimately associated with the rise of industrial society. These changes that permitted the bulk transport of goods and produce over long distances already lie at the root of many of the changes in rural employment over the course of the 20th century. Impacts have included the expansion of the geographic limits of production with which any particular economic activity has had to contend, increased competition and greater regional specialization in some activities. More recent changes in transportation technology, including containerization, have simply reinforced many of the patterns of change already underway. In some cases the result has been to strengthen particular regional economies, in others it has helped undermine them, particularly when they have been associated with 'consolidation and rationalization' processes in the shipping of agricultural produce. Transportation changes have, therefore, altered the boundaries of the systems of exchange that many rural activities function within.

The surge in developments in communications technology is perhaps the epitome of postindustrial society. The fax, telecommunications and extensive computer-based networks for the transfer of information have all contributed to the development of the 'global village' and the globalization generally of many aspects of the economic system – for instance, the globalization of capital. These changes are all information transfer or processing technologies. They typify the importance of knowledge as capital in today's society. On the one hand, they permit the decentralization of certain types of activities over huge distances; however, decentralization of the service activities that are dependent upon such technologies has not occurred to the extent that is physically possible. They have, however, had a significant impact on the evolving settlement structures and evolving employment patterns around many major metropolitan areas. The potential impact is even greater: telecommuting centres permit people to live in a decentralized settlement system and to go to 'work' several days a week in a local centre; the centre houses communication equipment and information processing equipment that a variety of companies and agencies might use or own collectively, thus eliminating the need to commute daily to a centralized office.

Many of the changes in production technology can also be traced back to the rise of industrial society. Most notable here was the development of massive scale economies in the production process of many sectors and the substitution of capital for labour. This form of technological change, together with the transportation developments that often went hand in hand with them, has underlain the massive changes in business structure in primary industries. Increases in 'optimal' business size in the face of relatively stagnant markets, together with the rise of competitors in other parts of the world, have led to plant closures and 'downsizing' of the labour force. Furthermore, the fact that, excepting for agriculture, many of the primary activities are internationally owned means that performance measurements in relation to capital invested in such enterprises are determined on an international basis, not a regional or even a national one. In agriculture, the changes have been no less dramatic, having led to widespread rural depopulation for many agricultural areas in North America and Western Europe.

On the other hand, while these same labour-saving technologies continue to play a significant role in some sectors, technological change in other sectors has permitted the growth and development of smaller and medium-sized enterprises. Examples include some sectors of the food-processing industry, where a combination of technology that favours (or at least does not penalize) small to medium size business and a more differentiated market for food products has encouraged diversification in production. Other examples include the development of hundreds of small desk-top publishing enterprises and other service-related activities where there is no need for the entrepreneur to come into face-to-face contact with the customer. While still largely concentrated in the major metropolitan areas, there are opportunities in these tendencies for some types of rural areas to derive benefits.

In areas where the rural economy is dominated by resource activities such as forestry and mining, rural economies have been subjected to downward pressures consequent upon technological change both in terms of production processes and in transportation, the accompanying globalization of markets and capital circuits, and corporate restructuring in relation to these. Agriculture has fared little better; this production sector is still highly fragmented but the nature of the markets faced by many farm areas has seemed to remove them from any control over their destinies.

These macro patterns and forces form the backdrop against which changes in rural economies and societies beyond metropolitan spheres of influence must be seen. They are also present in rural areas within the metropolitan regions; traditional rural activities there, especially farming, still represent important contributors to national income, even though absolute employment in these activities has decreased and other forms of employment have generally replaced this decline.

A Summary of the Chapters in Volume Two

The first two chapters deal with some of the key issues that arise from the processes of uneven development as our macro-social and economic structures are transformed. Then the discussion focuses successively on the transformation of social and demographic structures, the transformation of the use of space using the example of recreation and tourism and, finally, the transformation of economic structures.

Rural issues

Rural issues have both geographic and non-geographic dimensions. Both Cloke and Furuseth in Part I focus on the social consequences of uneven development, both in geographic and non-geographic terms. Cloke raises the issues of poverty and deprivation in rural areas, specifically in the US, linking them to economic restructuring processes and the maldistribution of income and jobs within communities. Some of the resulting out-migration of the impoverished has simply served to inflate the ranks of the 'urban poor', while a major response among those that remain in their rural origins is the heightened development of an informal economy. He argues that in relation to the 'issues', the notions of culture of poverty and underclass have been appropriated by a political discourse in the US in which the 'poor' are blamed for their impoverishment; this position reflects a lack of understanding of how the functioning of the broad system leads almost naturally to such extremes of uneven development. Furuseth's discussion of uneven development as a cumulative process, notwithstanding the short-lived period of the 'turnaround' in the US (also emphasized by Archer in Part II) calls attention to inequities in infrastructure provision such as in health care. He also argues that the new social and economic realities (the information society, the global economy, environmental awareness) all have a downside for disadvantaged rural areas, which highlights the need for effective local strategies to be developed, a theme reiterated in the contributions in Part V of this Volume.

The transformation of social and demographic structures

The broad geography of demographic change is tied directly to processes of uneven development – capital accumulation, geographic concentration and patterns of out-migration. Change in rural population and settlement is an almost continuous process, as Archer (Chapter 4) points out for the American 'Midlands'. The search for stable rural settlement structures, in the narrow sense of the term 'stable', is illusory. Archer illustrates this in relation to population in the American Midlands, and emphasizes the links between rural areas there and the external environment to which they are attached through various systems of exchange, such as world agricultural produce markets and technological change. He also notes the differentiation between rural areas, with those that showed consistently strong performances in demographic change being in or adjacent to major metropolitan areas.

At the community or locality level, uneven development can give rise to distinctive patterns of demographic change and social recomposition. Changes that are important include changing household size and socio-economic characteristics, resulting both from *in situ* demographic change and from differential movements of different population segments. Rural population ageing is one theme singled out for attention by Laws and Harper (Chapter 6). They emphasize rural population ageing as a complex process involving both *in situ* ageing and migration, with some areas developing as retirement zones for the relatively affluent elderly – thus emphasizing the heterogeneity of the rural elderly. Their pleas for the use of alternative research frameworks underlie a dissatisfaction with the positivist approach which has given rise to relatively narrow interpretations and descriptions of rural ageing.

Attendant upon changes in demographic structure and social recomposition are changes relating to changing needs of the population in terms of physical and social infrastructure. Medical services and social services are emphasized by Furuseth in Part I and by Laws and Harper in Part II. Inequities in access to housing are developed in detail by Robinson (Chapter 7) for the UK. His discussion of housing supply highlights differences between rural areas and between rural residents; he notes especially the difficulties that long-time rural residents must face in some areas due to the competition from an influx of urban residents into parts of the countryside for permanent residence (including retirement) and for second homes. These difficulties have been exacerbated in the UK since the late 1970s as new policies were developed to permit the sale of council (public sector) housing, giving rise to lengthy waiting lists for the remaining council housing stock. Responses have included the increasing role and support for housing associations and trusts in providing (low-cost) housing, as well as special negotiated arrangements between local authorities and developers. The future may well include an even greater development of 'self-help' and locally initiated responses.

Finally, as rural communities have become transformed in relation to the broad changes occurring in the macro and regional environments in which they function, various social stresses are created within the community, some of which are noted by Cloke (Chapter 2) Furuseth (Chapter 1), Laws and Harper (Chapter 6) and Robinson (Chapter 7). In some cases, these stresses are related to the economic restructuring of activities in rural areas, for example agricultural rationalization and agricultural decline. Mackenzie (Chapter 5) discusses in particular the situation and the response of one segment of the rural population, farm women, and how in Eastern Ontario the farm crisis initially deepened the feeling of powerlessness, both of the farm population generally and of farm women in particular. This led to an organized response through a network of farm women – Farm Women for the Survival of Agriculture – who have blended a nationalist ideology, in which the family farm becomes identified with the survival of Canadian sovereignty, and a feminist ideology, which seeks to challenge and restructure the 'farmer and his wife' hierarchical relationship in farm life.

In other areas, the stresses stem partly from injection of new 'blood' into rural communities, particularly in the context of rural communities lying in the metropolitan field. Here the centrifugal forces of metropolitan decentralization have brought different groups of people into contact, for instance the farming population and various newcomers. Bunce and Walker (Chapter 3) focus on the property system and the land market as the vehicles through which these changes have occurred in Toronto's countryside. The combined processes of exurbanization, bringing in exogenous capital investment by people seeking to 'buy' the amenity of the countryside, and of ruralization, related to the restructuring of agriculture, has given rise to a variety of settlement changes and to a heterogeneous population structure. They underscore the fact that many negative changes in agriculture are not related to exurbanization but rather to ruralization processes linked to relatively low returns in farming.

The transformation of the use of rural space

The transformation of social, demographic and economic structures is inevitably linked to changing patterns of use of 'rural' space. This occurs at a variety of scales, ranging from the use of the home as a business base (Dykeman, Chapter 19), to changing patterns of housing need and supply (Robinson, Chapter 7), to changing agricultural production structures (see Volume 1), to changing patterns of economic development (see following section), and to changing patterns of usage and valuation of rural space by people who are not resident in the rural areas concerned, notably through recreation and tourism (Part III).

Recreation and tourism in rural areas have given rise to a fairly extensive literature. Increasing attention over the last 20 years partly reflects the changing value patterns of recreationists and tourists for access to rural recreational and tourism opportunities. This in turn reflects increasing concerns for individual health and a mounting interest in 'nature' and 'environment'. Mason (Chapter 8) uses a discourse similar in many ways to that of Bunce and Walker's. He focuses on the protection of regional environmentally-valued landscapes through land trusts in the US and argues that, in many cases, this movement reflects an elitist view of landscape by people trying to protect their image of the rural idyll. Not surprisingly, much of the activity Mason describes is occurring in the urban and suburban fringes of the major metropolitan centres of the US – areas where there are both pressures for change and a population concerned with protecting 'its' own environment.

Recreation and public access can present conflicts with environmentally and aesthetically valued lands. Debate over the merits of the integration of recreation and access, on the one hand, and conservation, on the other, must be placed in the context of value sets and power relationships. For the UK, Curry (Chapter 9) argues that recreation and access deal with *public* enjoyment of open space and attractive environments, and that this has been in continual conflict with conservation values (aesthetic or scientific), with the conservation values winning out in terms of policies in which recreation and access are residualized. Furthermore, much of the debate regarding recreation in rural areas is undertaken from the perspective of the urban user, and Glyptis (Chapter 10) seeks to provide some balance by considering the needs and expectations of rural residents. Reporting on a wide range of studies in very different rural areas, Glyptis finds very little evidence of massive deprivation, although specific types of rural residents do confront access problems, namely, women, young people and retired people, especially when combined with lack of car ownership.

Butler and Clark (Chapter 11) extend the discussion to include rural tourism in the UK and Canada. Although there are important differences between the two countries, particularly in terms of the lack of a coherent set of images of rural areas in Canada, the issues and concerns raised are common. Rural recreation and tourism are generally viewed as expanding and as being a major avenue for economic development of rural areas (this discussion therefore overlaps partly with Part V). However, the authors point out the necessity to undertake a critical analysis of the costs and benefits of rural tourism – economically, environmentally and socially. 'Success' also brings its own problems, because (economic) success can easily destroy the very values that supported the growth of rural tourism in an area. Hence, difficult issues are raised again for management and planning for tourism, because this brings us back full circle to comments made by Mason (Chapter 8) and Curry (Chapter 9) regarding conservation (protection) representing an elitist set of values.

The transformation of economic structures

Economic restructuring underlies many of the issues and concerns associated with the transformation of rural areas and their changing demographic and social composition. Many of the broad processes have been noted earlier in this Introduction, as well as in some of the contributions to Volume 1 in terms of agriculture and agribusiness. They have also been dealt with at length in two recent books (Lowe *et al.*, 1990; Marsden *et al.*, 1990).

A number of phenomena linked to these various forces of change lie at the root of the transformation of economic structures. Changing market structures and competition related to globalization and corporate restructuring internationally have led to significant changes that have affected rural areas differentially. Several contributions in Volume 1 discussed the ramifications of these changes. Changing values are powerful factors operating through the market system, as well as influencing governmental intervention in a number of processes. The increased value attached to enriched personal experiences and personal health are partly behind the interest shown today in rural recreation and tourism (Butler and Clark, Curry, Glyptis, Mason); while increased concerns regarding personal health and the environment are linked to changing markets for agricultural produce (see Volume 1) as well as for other rural products. The main feature of these changes seems to be the increasing link between process and product in the sale of the product, providing a critical link to environmental sustainability.

Changing technology, together with changing market patterns and the increasing importance of service activities, have led to important changes in business development patterns. In particular, the greater emphasis on small-scale business development has opened up greater possibilities for economic development in small rural communities because of the greater degree of compatibility between such enterprises and the social and economic stability of the community. One important question receiving more attention is how such small-scale businesses get started. Many originate in a home base because of the lower overhead costs involved and the greater flexibility in running such an activity on a part-time basis in the early phases. Dykeman (Chapter 19) draws upon extensive evidence for Atlantic Canada to show how important this phenomenon is in rural areas; he highlights the valuable contribution such activities make to rural economies and life-styles, even though home-based businesses have often not been treated seriously as economic development tools.

Similarly, economic restructuring has brought with it significant changes in rural employment patterns in the primary industries. This has been emphasized nowhere as much as for the agricultural sector. In Part IV, Walford (Chapter 12) highlights the effects of restructuring on the agricultural workforce, an employment group that has more often than not been ignored in studies of rural employment change. He emphasizes the decline in the full-time agricultural worker and an increase in seasonal and temporary workers in the UK. Fuller and Bollman (Chapter 13) discuss multiple job-holding in farm families, drawing upon an extensive research base in a dozen countries. They argue that rural labour markets are restructuring in relation to macro-scale forces. Agricultural production is becoming 'decoupled' from traditional support modes (subsidies and guarantees linked to production) and the farming population is being encouraged to turn to other uses of farm and family resources. Part of this adjustment is the growth of pluriactivity, both in North America and Europe. Overall, they conclude that pluriactivity is not only the result of agricultural restructuring but also of social restructuring, reflecting the greater participation of women in the work force, increased preferences for part-time work as a life-style and the decline of single-career paths.

The responses to change

Responses to the transformation of rural societies and economies are multi-faceted and occur at several levels. At the individual, household and farm or firm level, the transformations themselves represent the collective response of these elemental decision-making units to the forces of change. Sometimes group or collective action can rise to cope with and initiate change, for example Mackenzie's discussion of farm women in Eastern Ontario; response can also occur at the community (local or regional) and 'state' level (state, province, federal government).

There are many types of responses to the forces affecting rural areas. Three broad categories of response, particularly at the individual or community level, are: winding down behaviour (for example, preparing for or taking retirement, preparing to quit the activity, falling into a situation of dependency upon state or other forms of collective help), coping behaviour, and adapting and innovating behaviour. In relation to adapting and innovating behaviour versus winding down behaviour leading to dependency, it is interesting to note that the state can create a favourable enabling environment or one that is unfavourable to such action (Bryant, 1989).

In this Volume, several chapters stress the importance of individual responses and adaptations, including multiple job-holding (Fuller and

Bollman), and the emergence of home-based business as a force to reckon with in rural economic development (Dykeman). At the community level, which is clearly a major preoccupation of the Canadian and US contributions, a number of responses and strategies are discussed. Increased demands for greater local and community responsibility over managing and planning change are major themes for Bryant (Chapter 18) who sees greater community involvement in economic development through strategic planning, with a trend away from the traditional 'industrial development' model towards more community-oriented approaches. Wolfe (Chapter 20) deals with a strongly related domain, the issue of self-governance for aboriginal groups in Canada. In a more general vein, Everitt and Annis (Chapter 14) argue that a greater weight be placed on intervention to support rural communities because of their overall value in the broader economic and social system. They argue that the 'economic deck' is stacked against the small town in Canada and that 'unstacking' the deck is necessary. Specific local strategies are singled out by some authors, for instance the encouragement of home-based business (Dykeman, Chapter 19) and the local provision of workshops in the UK to help small business start and grow (Martin, Chapter 17). Interestingly enough, one of the results of Martin's study in the English Midlands shows that the impact on rural employment was not so much to reduce 'registered' unemployment but rather to 'soak up' concealed unemployment. This leads Martin to make a plea for more research on the functioning of rural labour markets.

State intervention occurs in many forms, ranging from direct support of a charitable nature (Cloke, Chapter 2) to creating favourable enabling environments that encourage community initiative (Bryant, Chapter 18). There are also more direct forms of intervention aimed at enticing or encouraging the development of different economic activities in rural areas. However, it is clear that the expected effects of such 'state' efforts have often proved difficult to attain and that other non-economic impacts have not always been taken into account adequately. Pyle (Chapter 16), for instance, discusses federal water resource development in the US that has frequently been justified as creating jobs and regional economic development. Reservoir construction was frequently the outcome of effective 'pork-barreling' in the post-war period, and can be viewed as akin to 'smoke-stack' chasing by local and regional representatives. However, Pyle questions the ability of these projects to generate employment and also criticizes them for paying too little attention to other values, including environmental and amenity values.

Clearly, any form of intervention to encourage economic development raises the question of whether the intervention is really oriented towards a truly sustainable development process, in which a wide variety of values and interests have been taken into account. Inevitably, the debate has been heightened over potential conflicts between environmental values and jobs through economic development, particularly in terms of rural and small town environments, which frequently still lag behind metropolitan and urban places in economic prosperity. Frederic's (Chapter 15) analysis of economic development and environmental regulations for most of the US states shows a negative relationship between the number of economic development support options at the state level and environmental controls. He suggests, however, that development programmes seem to be ineffective and that environmental regulations do not prevent growth in any case, being more of a product of growth than anything else. He concludes with a plea for setting aside the dichotomy between development and environment and paying more attention to integration. This plea for the integration of values in a broader developmental process undoubtedly gives us an important clue for directions in both future research and policy.

References

Bryant, C.R. (1989) Entrepreneurs in the rural environment. *Journal of Rural Studies* 5, 337–448.

Lowe, P., Marsden, T. and Whatmore, S. (eds) (1990) *Critical Perspectives on Rural Change Series. Vol. 2: Technological Change and the Rural Environment.* David Fulton, London.

Malassis, M. (1958) *Économie des exploitations agricoles: essai sur les structures et les résultats des exploitations agricoles de grande et de petite superficie.* École Pratique des Hautes Êtudes, Paris.

Marsden, T., Lowe, P. and Whatmore, S. (eds) (1990) *Critical Perspectives on Rural Change Series. Vol. 1: Rural Restructuring: Global Processes and their Responses.* David Fulton, London.

Naisbitt, J. (1982) *Megatrends.* Warner Books, New York.

I

IDENTIFYING THE ISSUES FOR THE RURAL ECONOMY AND SOCIETY

1

UNEVEN SOCIAL AND ECONOMIC DEVELOPMENT

Owen Furuseth

Despite a rural heritage, recent US government policy toward rural areas has been mixed and often misguided. The reasons for this failure are complex and in some cases interrelated. Sher (1986) has observed that American rural policy tends to vacillate between alternative visions of the countryside: the bucolic and the bubonic. In periods such as the Reagan era, the bucolic imagery prevailed. Within this genre, rural areas are the Jeffersonian ideal: an agrarian paradise, where serious economic and social problems are absent. The alternative bubonic vision sees rural areas as derelict landscapes, populated by the leftover disadvantaged persons. Lyndon Johnson's 'Great Society', for example, focused a great deal of attention and effort on the plight of the rural poor. Although elements of both the bucolic and bubonic perspectives are evident on the rural landscape, neither is singularly appropriate for describing rural America.

Uneven Social Development

A more salient perspective on rural America in the last two decades of the 20th century is uneven social development. The notion of unevenness in social relations represents an extension of the concept of uneven economic development. In general, the theory of uneven development is used to conceptualize the differing rates of industrialization or economic growth in capitalist economies (Malizia, 1978; Bluestone and Harrison, 1982; Gordon et al., 1982). According to Bluestone and Harrison (1982), the increasing concentration of capital and a concurrent inability of labour to move freely encourage the uneven growth of industries, regions and social groups. The most serious consequence of this process is increasing inequalities that cannot be corrected by the free market.

Analysts find two major reasons for uneven development. First, the initial capital investment process alters the quality of regions, labour and industries. Initial successes, in turn, foster increasing success in the future. For example, increased investments in infrastructure, human resources, or product quality pay off with increased opportunities in the market; conversely, opportunities shrink for others who cannot or do not make such investments. The cumulative effects of early successes or failures in markets reinforce the polarization of the 'winners and losers' (Colclough, 1988).

A second explanation for uneven development derives from the perspective of the advantaged. Once one has gained economic advantage there is a keen interest in maintaining or strengthening that advantage. The uneven development process dictates that winning industries, regions or groups will attempt to continue enhancing their positions; and as capital concentration grows, so does inequality.

Uneven social development is a consequence of economic inequity. It reflects inequality in access to human resources between regions and/or populations. In the US, where the State has no obligation to intervene seriously and balance social inequalities, differentials in education, health care, housing and environmental services aggregate to shape pronounced differences in the social fabric of a community. Isolated from the mainstream of society, these areas are persistent laggards in measures of socioeconomic welfare. Cloke (Chapter 2), observes that a distinctive underclass culture is found in portions of rural America. According to Cloke, concepts like hopelessness and disadvantage, traditionally associated with inner city ghettos, are equally salient in many rural locales. Beyond immediate affects on the disadvantaged community, social inequalities contribute to economic disparity. Uneven social and economic inequalities are intertwined in a vicious cycle. Economic differentials trigger social disparity, which leads, in turn, to further economic differentials. Without structural change, the cycle continues unabated.

This chapter examines rural social conditions in the US during the past decade, with an assessment of likely conditions in the future. This period has been marked by a significant economic restructuring in the global economy which has given rise to major impacts on the American countryside. The events of the past decade have accelerated the pattern of uneven social development between rural and urban areas in the US. More significant, however, is evidence of a growing uneven development within rural America. While some rural communities have benefited greatly from the restructuring process, other rural communities have been irreparably affected. This paper suggests that the current pattern of uneven social development begun in the 1980s will continue through the 1990s, resulting in even further differentiation in rural America by the end of the century.

Economic Restructuring in the 1980s

Global economic events of the 1980s caused significant changes in the US economy throughout the decade. While all sectors of the US economy and all regions felt the impacts of economic downturn and recession, rural areas and population were especially vulnerable. Ironically, the events of the 1980s followed a decade of dramatic rural prosperity. The 1970s had been heralded as the 'rural renaissance', an economic and demographic renewal creating optimism about the future of rural communities. In particular, the 'industrial invasion' of non-metropolitan areas during the 1970s was seen as a harbinger of long-term economic growth, and the emergence of a strong, seemingly sustained, social revitalization for rural communities (Summers *et al.*, 1976).

The heady times, however, quickly ended at the beginning of the 1980s. The population and economic growth of the preceding decade turned out to be ephemeral. Economic decline and depopulation re-emerged as primary concerns of rural America. During this decade, the simultaneous loss of low-wage manufacturing jobs to overseas locations, the farm recession, and the downturn in energy industries combined to create widespread rural decline. Economic gains made by rural areas, relative to metropolitan areas, in the 1970s were lost. While job growth in metropolitan areas during the 1980s simply slowed down, rural areas experienced job losses. For example, rural unemployment increased from 5.7% in 1979 to 10.1% in 1982 (Bloomquist, 1987). In the early 1980s, the average rural unemployment rate was 7% greater than the urban rate; and by 1987 it was 40% higher.

Social Consequences of Restructuring

Not unexpectedly, the economic distress of the 1980s led to a deterioration in rural social conditions. Poverty rates in non-metropolitan areas, which had been declining since 1962, began increasing in 1979 (Porter, 1989). They continued to rise until 1986, since when they have remained fairly constant. By 1987 one out of every six rural families lived in poverty (US Department of Commerce, Bureau of the Census, 1989). The most recent data (1988) indicate that non-metropolitan poverty was 3.8% above the metropolitan rate (US Congress, Office of Technology Assessment, 1991). The scale of the problem is particularly serious for minorities and disadvantaged populations: African Americans, Native peoples and Hispanics.

The economic restructuring of the 1980s was especially devastating for the rural poor because of their precarious economic condition. Most rural residents living at or below the poverty level are 'working poor', that is, they are employed in low-wage jobs (Porter, 1989). In many instances these families will have more than one wage earner. For these individuals the loss of any income, whether because of short-time layoffs or permanent job loss, affects their ability to sustain economic and social survival.

As would be expected, rural social infrastructure struggled during the period. In broad historic terms rural America has been poorer, less educated, poorly housed, and more likely to suffer health problems than urban America. During the 1980s all of these socioeconomic indicators showed an erosion of already marginal rural conditions against urban America. Not only were rural residents more likely to be unemployed but their communities lacked the resources to provide support services comparable with those available to urbanites. Accordingly, the social impact of the 1980s economic distress was harder in places like rural Iowa and Georgia than in Detroit or Houston.

Restructuring and the Poorest Areas

While all rural America felt the impacts of the 1980s, the poorest, most disadvantaged areas were especially hard hit. The Southeastern US contains 25 million of the 57 million rural poor; 40% of rural Southerners are either poor, elderly or both. Indeed nowhere in the US was the difference in impact caused by the 1980s restructuring more evident than in the South. In the 'sunbelt' states a sharp, rural-urban dichotomy developed (Fig. 1.1). Generally the MSA (Metropolitan Statistical Areas) counties and other urban counties, cushioned by trades and service industries, avoided economic and social disruption, whereas rural counties dependent on low-wage, low-skilled jobs became the backwaters of the American economy (Falk and Lyson, 1988).

For state and local government leaders used to 'selling' the rural south based on a package of cheap and willing labour, weak government regulations, as well as financial inducements, the new global economic arrangements have been perplexing. During the 1960s and 1970s, Southern industrial recruiters methodically lured standardized manufacturing jobs from Northern urban areas. During the 1980s many of these jobs were lost to lower-waged international manufacturers and relocated outside the US. As Rosenfeld (1983) notes:

> . . . rural communities, which have been primed for 1960s-style
> growth, with waste disposal systems, roads, industrial parks and
> vocational-technical centers, now find themselves facing a new
> 1980s-style growth. Just as they began to catch up, the ground
> rules have changed. A skilled work force, strong schools and
> extensive communications links have been added to the list of
> factors needed to sustain growth.

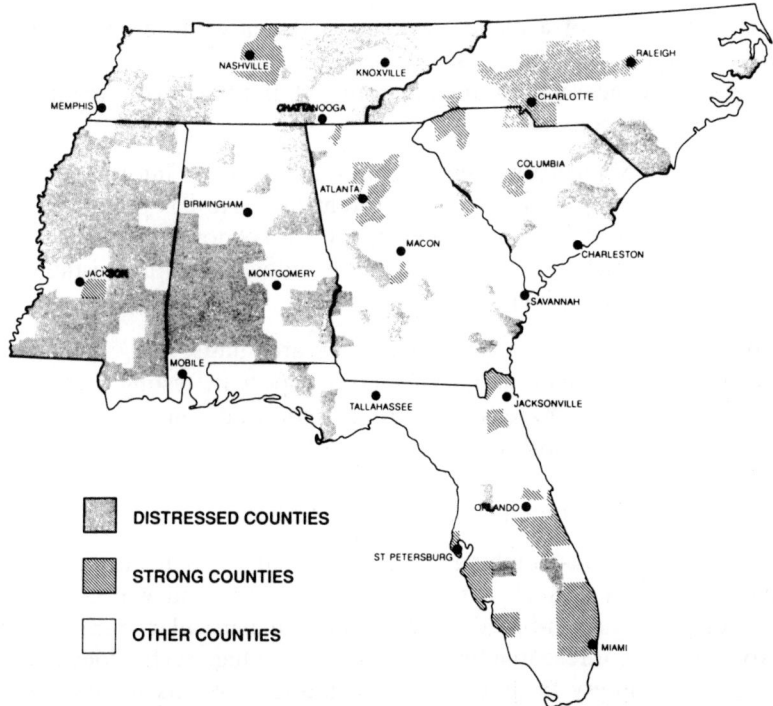

Fig. 1.1. Distressed and strong counties in the southeast. Source: After Aiken, 1989.

For many rural southern communities, especially the poorest and those with concentrations of African Americans, the impacts of the 1980s restructuring have not ended. They remain trapped by the effects of past economic policies and the current recessionary milieu. Most importantly, the future for these communities holds limited opportunities.

Beyond the 1980s

The current changes in global economies continue to affect rural economic and social conditions. How well rural areas are able to cope with the on-going shifts is a function of their present situation and resources and the role that the state may play in helping adapt to these changes. A report by the US Congress, Office of Technology Assessment (1991), warned that there are three major trends likely to have profound effects on rural communities in the near future. They are:

1. the shift to an information-based economy and the enhanced role of communication and information as a strategic weapon in business;

2. the emergence of a global economy and hence the growing need to compete on a worldwide basis;
3. a growing concern about the environment and the environmental costs of economic development.

Castells (1985) has noted that the present shift to an information-based economy is shaping a new space of production. He has proposed a new locational decision-making model for high-technology manufacturing. Using this approach the combination of the following characteristics makes an area attractive for high-tech activities:

1. location near major universities and research units;
2. location in or around military facilities, especially testing sites;
3. location away from areas with strong union traditions;
4. location in areas with concentrations of or access to venture capital; and
5. good positions within a communications network.

The serendipitous combination of locational attributes has positioned some rural areas in such a fashion that they are and will continue to enjoy the information-based restructuring. In particular, rural counties surrounding major research universities, existing high-technology research centres and military facilities have become focal points for civilian and military-based technological research and manufacturing. Similarly, a number of isolated rural communities in the Western US, including Oregon, Arizona, Colorado and Texas have been 'discovered' by high-technology firms and transformed into high-technology production nests (Castells, 1985).

The corollary is, of course, rural areas without these locational attributes. Add to this formula pre-existing disadvantages, such as a lack of educational resources, inadequate infrastructure, poor health care and racism, and it raises the question what does the information-based future hold for these rural areas?

Falk and Lyson's (1988) analysis of high-technology employment in the South foresees serious economic problems for these rural areas. They found tremendous growth in the number of high-tech jobs in the region; between 1977–1981, 400 000 new jobs were created. Accordingly, by 1981, nearly 9% of all private sector employment in the South could be identified as high technology. Nevertheless, most of these jobs were located in urban areas and very few went to the poorest rural areas. Over 87% of new jobs were in metropolitan counties. While 'Black belt' counties, areas with concentrations of African Americans, garnered only 0.04% of these jobs. Most recently, the high-technology gap between the urban and rural south has continued to grow wider.

The second trend noted in the *Crossroads* report – the globalization of the economy – is actually a continuation of the economic restructuring begun in the 1980s. The hallmark of this period was increased trading competition between American and international firms. The decline in the US trading position fell hardest on rural communities. For example, between 1975–1982 an estimated 11 000 rural jobs were lost to imports (US Congress, Office of Technology Assessment, 1991). In contrast, unemployment in metropolitan areas did not increase due to imports. During this time, 201 rural counties experienced a loss of more than 10% of their manufacturing jobs.

If rural areas are to survive in the increasingly globalized economy then they must become globally oriented. In other words, they need to be able to attract international and domestic investment focused upon the global marketplace. They also need to develop their local entrepreneurial base. Locally-based small firms providing manufacturing and service jobs can play a significant role in rural community development.

The obstacles to making the global transition are many of the same issues affecting the shift to the high-technology, information-based economy: underdeveloped human resources, shortage of investment capital, racism, and isolation from entrepreneurial and scientific innovation. Certainly, some rural areas have positioned themselves well to make the transition to global restructuring. Those that have invested in human resources, or are located near metropolitan areas, are particularly suited for globalization. But for other areas, including the traditionally poorest rural counties, which have focused their economic plans on low-wage, standard manufacturing, the barriers are significant.

The third trend reflects the growing sensitivity to the environmental cost of economic growth and a new environmental paradigm among the US population. This paradigm is carried over into the business location decision-making arena. Appealing physical settings and/or positive environmentally related perceptions have made some rural areas attractive, particularly for knowledge-based industries (Myers, 1988). Thus rural communities located in Northern New England or the Intermountain West may be geographically isolated, but their pristine natural environment engenders a strong quality of life appeal to business interests. Unhappily for rural leaders, not all rural communities enjoy this special aura and appeal.

For a growing number of rural communities, environmentalism has provided a more insidious set of economic opportunities. Rural areas are increasingly becoming the hosts for unwanted and less desirable economic ventures and land users. In the face of rising NIMBY-ism (Not in my back yard) among urban, suburban and wealthier rural areas, growing amounts of municipal waste, toxic and hazardous materials, and nuclear waste materials are being disposed of in rural areas. 'Trash for cash' wastes are often shipped hundreds of miles to isolated, often disadvantaged, communities. In 1983, the US General Accounting Office documented the correlation between hazardous waste disposal sites and rural Southern black populations; Bullard (1990) and others have further developed this relationship.

Recently, the US Environmental Protection Agency and US Department of Energy were chastized for joining with municipal and toxic waste disposal firms in coercing Native American populations to accept waste disposal facilities on their reservations (Tye, 1991). The Office of the US Nuclear Waste Negotiator has written to the nearly 300 Indian tribes announcing the opportunity to become a repository for high-level nuclear waste (Lambrecht, 1991). Faced with enormous social and economic handicaps, and promised economic rewards, waste disposal proposals are very appealing to native peoples. Empirical data suggest that Indian reservations are an increasingly attractive locational target for waste disposers (Kay, 1991). Writing about 'environmental racism', Lambrecht stated:

> Waste merchants look to remote, self-governing Indian territory partly because state and local environmental laws do not apply there. Until recently, they found tribal councils easier to deal with than other governments.
>
> A pattern is developing. Indian leaders quietly sign million-dollar contracts. Then they face uprisings from tribal members who insist that Indians lack the know-how to monitor wastes sites or even to evaluate proposals.
>
> Typically, rancorous debates break out over trading land that is revered for the promise of cash payments, scholarships, and better health care.
>
> (Lambrecht, 1991)

The Future

The events of the past decade combined with expected changes present a variety of challenges for rural communities. The most recent national statistical data together with the evidence from empirical studies indicates increasing uneven social development within rural America. Thus, there is reason to suggest that the growing disparity between 'have' and 'have-not' rural areas of the past decade will continue.

For many rural jurisdictions the economic restructuring environment of the 1980s presented enormous opportunities for improving socioeconomic conditions. For the most part, these counties were already among the most advantaged rural communities in the US. The events of the 1980s provided the potential to increase further their uneven position relative to poorer rural jurisdictions, and to allow them to evolve socially into more urban-liked communities.

Those rural counties that sustained growth in the 1980s are most likely to continue to grow in the 1990s. Analyses have shown that several distinctive social, demographic and economic attributes marked these areas (MDC, 1986). They were counties that had the smallest percentage of minority populations, the highest educational levels and per capita income, more physicians per resident, and large numbers of in-migrating elderly (retirees). Economically, these rural areas had economies focused on service industries, rather than manufacturing, agriculture or resource activities. Locational elements were also identifiable: access to interstate highways and proximity to metropolitan areas were associated with high growth.

For geographers, one of the most interesting factors at work in this process is proximity to urban centres and access to interstate highways. Rural areas near metropolitan areas, or connected easily to metro areas by high-speed highways, continue to be socially and economically impacted by their urban neighbours. The social convergence between rural and urban is most active and pronounced in these areas, with growing urban influences on these communities. Unfortunately, the urban influences include not only the positive elements – the high-tech jobs, the increased availability of medical services, and larger school budgets – but also the negative impacts – increased violent crimes, drug abuse, AIDS and other social problems. In a real sense, the challenges posed by urban convergence may be almost as serious as the problems facing disadvantaged rural areas.

For these latter communities, the future is clouded by the on-going restructuring as well as the legacy of disadvantagement. Two groups of counties are particularly vulnerable to falling further behind other rural areas: they are those labelled by Davis (1979) and Hoppe (1985) as persistently low-income (PLI) counties and those that are racially identifiable. Most are located in the South (Fig. 1.2). Because of existing conditions and racism, these communities cannot effectively respond to the restructuring conditions. Their role in the changing rural scene is increasingly marginal. White (1989) has argued that Appalachian Kentucky has become America's Soweto – a ready source of labour for urban-based manufacturing centres. Bullard (1990) has warned that low-income African-American communities are at risk of becoming dumping grounds for toxic and hazardous wastes. Evidence of environmental racism in Native American communities is also growing and some policy analysts suggest that government practices 'triage'-style intervention (Malizia, 1985). Thus only rural areas that have a strong potential for economic and social recovery warrant state assistance; weaker, less promising rural areas would be unassisted. Whatever the course of action (or inaction), the immediate future for

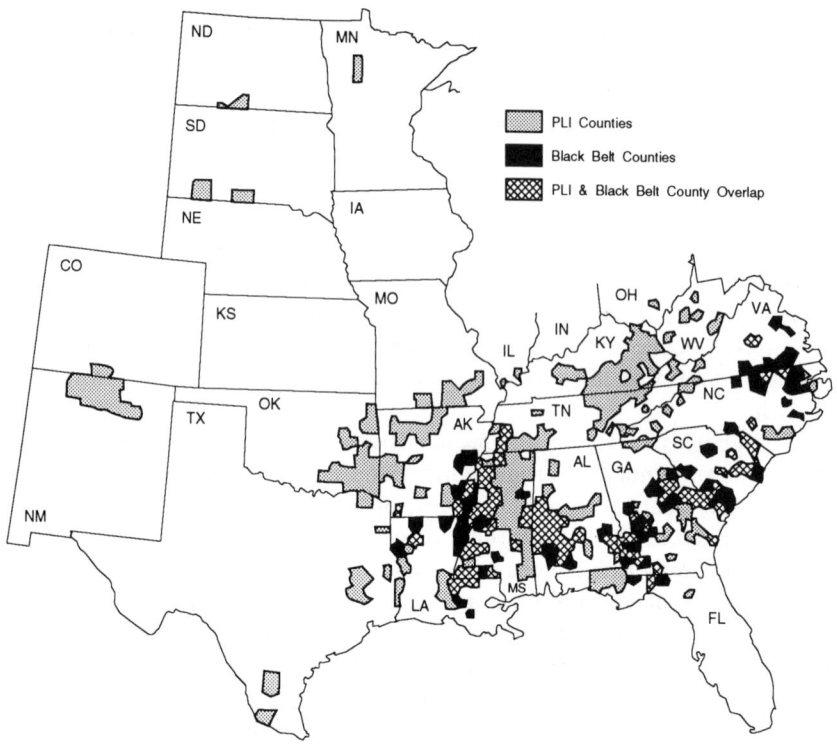

Fig. 1.2. Disadvantaged rural areas.

these rural residents and communities labelled 'backwaters of American economy' does not look promising, as uneven development continues its course.

References

Aiken, C. (1989) The Rural South as seen by a historical geographer. In: Castle, E. (ed.), *National Rural Studies Committee, a Proceedings*. Western Rural Development Center, Corvallis, OR, pp. 11–22.

Bloomquist, L.E. (1987) Performance of the rural manufacturing sector. In: *Rural Economic Development in the 1980s Preparing for the Future*. Agriculture and Rural Economy Division, Economic Research Service, US Department of Agriculture, US Government Printing Office, Washington DC, pp. 3-1–3-33.

Bluestone, B. and Harrison, B. (1982) *The Deindustrialization of America*, Basic Books, New York.

Bullard, R.D. (1990) *Dumping In Dixie, Race, Class and Environmental Quality*, Westview Press, Boulder.

Castells, M. (1985) High technology, economic restructuring and the urban-regional process in the United States. In: Castells, M. (ed.) *High Technology, Space and Society*, Sage, Beverly Hills, pp. 11–40.

Colclough, G. (1988) Uneven development and racial composition in the Deep South: 1970–1980. *Rural Sociology* 53, 73–86.

Davis, T.F. (1979) *Persistent Low-Income Counties in Nonmetro America*. Rural Development Research Report 12, Agricultural Economics Statistical Cooperative Service, US Department of Agriculture, Washington DC.

Falk, W.W. and Lyson, T.A. (1988) *Hi Tech, Low Tech, No Tech: Recent Industrial and Occupational Change in the South*. State University of New York Press, Albany.

Gordon, D., Edwards, R. and Reich, M. (1982) *Segmented Work, Divided Workers: The Historical Transformation of Labor in the United States*. Cambridge University Press, New York.

Hoppe, R.A. (1985) *Economic Structure and Change in Persistently Low-Income Nonmetro Counties*. Rural Development Research Report 50, Economic Research Service, US Department of Agriculture, Washington DC.

Kay, J. (1991) Indian lands targeted for waste disposal sites. *San Francisco Examiner*, 10 April, A–10.

Lambrecht, B. (1991) Tribes asked to take nuclear waste. *St. Louis Post Dispatch* 28 April, A–1.

Malizia, E. (1978) Organizing to overcome uneven development: the case of the US South. *The Review of Radical Economics* 10, 87–94.

Malizia, E. (1985) *Economic Growth and Change in the Nonmetropolitan South*. MDC Inc., Chapel Hill.

MDC (1986) *Shadows in the Sunbelt*. MDC Inc., Chapel Hill.

Myers, D. (1988) Building knowledge about quality of life for urban planning. *American Planning Association Journal* 55, 357–73.

Porter, K.H. (1989) *Poverty in Rural America: A National Overview*. Center on Budget and Policy Priorities, Washington DC.

Rosenfeld, S.A. (1983) Prospects for economic growth in the nonmetropolitan South. *USGPB Alert*, p. 4.

Sher, J. (1986) Rural development worthy of the name. In: US Congress Subcommittee on Agriculture and Transportation, Joint Economic Committee *New Dimension in Rural Policy: Building Upon Our Heritage*. US Government Printing Office, Washington DC, pp. 515–16.

Summers, G.F., Evans, S.D., Clemente, F., Beck, E.M. and Minkoff, J. (1976) *The Industrial Invasion of Nonmetropolitan America. A Quarter Century of Experience*. Praeger, New York.

Tye, L. (1991) Indians rebel against waste dumps on reservations. *Wilmington Star*, 29 June, A–1.

US Congress, Office of Technology Assessment (1991) *Rural America at the Crossroads: Networking for the Future*. US Government Printing Office, Washington DC.

US Department of Commerce, Bureau of the Census (1989) *Rural and Rural Farm Population: 1988*. US Government Printing Office, Washington DC.

US General Accounting Office (1983) *Siting of Hazardous Waste Landfills and Their Correlation with Racial and Economic Status of the Surrounding Communities*. US General Accounting Office, Washington DC.

White, S.E. (1989) America's Soweto: population redistribution in Appalachian Kentucky, 1940–1986. *Appalachian Journal* 16, 350–70.

2

RURAL POVERTY: SOME INITIAL THOUGHTS ON CULTURE AND THE UNDERCLASS

Paul Cloke

The Research Context in Britain

Despite the recent emphasis on social recomposition in rural Britain, very little is actually known about the extent of poverty and deprivation in rural areas. A number of recent publications (Glendinning and Millar, 1987; Atkinson, 1989; Andrews and Jacobs, 1990; Oppenheim, 1990; Barooah *et al.*, 1991; Brown and Scase, 1991) suggest a resurgence of interest in poverty at a national scale, but much of our understanding of contemporary poverty in rural areas relates back to the early 1980s when the last major research initiatives occurred. The conceptualization of deprivation at that time is excellently reviewed by Lowe *et al.* (1986), and empirical evidence has largely been drawn from the research project undertaken by McLaughlin and Bradley for the Department of the Environment (McLaughlin, 1986).

Because this study is the only one of its kind in rural Britain in the 1970s and 1980s, the results that emerged have received considerable attention. Intensive surveys were undertaken of households in five areas of rural England, and the research was complex and lengthy (an 80-page questionnaire!) such that no one particular factor can summarize adequately the multi-faced nature of the findings of the study. Nevertheless, one indicator did catch the imagination. McLaughlin and Bradley generated an index for each of the five case-study areas in which annual gross disposable income was expressed as a percentage of the supplementary benefit scale rates, plus actual housing costs, for each household. A threshold for 'deprivation' was then fixed for this index:

> using this calculation, households with incomes of up to 139% of their supplementary benefit entitlement are identified as living in or near the margins of poverty.
>
> (McLaughlin, 1986, p. 294)

The following percentages of households at or below the 139% threshold were found for each of the five case-study areas: Essex 24.9%, Yorkshire 25.8%, Suffolk 21.4%, Shropshire 24.9%, Northumberland 27.3%. These figures have been seen as remarkable not only for their high level, but also for their apparent consistency across widely differing parts of Britain.

To some extent this evidence has become a conventional wisdom in rural studies in Britain. It has become increasingly clear, however, that there are many issues left unresolved by this work:

1. What is the socioeconomic anatomy of 'the 25%' deprived? What are the interconnectivities between deprivation and class, gender, age, ethnicity and localism?

2. Are there different geographies of deprivation? Are the Essex '25%' really the same as the Yorkshire '25%' or do these seemingly similar results mask important localized factors (state action, labour markets etc.)?

3. What about those forms of deprivation that are not dealt with by a single economic indicator? How important, for example, are *cultural* privations of choice, access, opportunity, security etc.?

4. How significant is it that some rural respondents to the surveys acknowledged their own experience of deprivation, whereas for many others such experience was unacknowledged even though it was measured normatively.

5. How has the position changed since 1980 during the era of the Thatcher economic miracle (Cloke, 1992)? After all, John Moore, Secretary of State for Social Security in 1989, claimed that economic success had put an end to absolute poverty in Britain.

Given these uncertainties, it can be argued very strongly that there is a need for considerable conceptual discourse as well as for further empirical evidence. It is on this basis that a major research initiative is now underway to study rural life-styles in Britain, in which detailed surveys are being undertaken in 16 areas of England and Wales (including the five originally studied by McLaughlin and Bradley). So as to inform that study, a prior examination has been made of poverty in the United States.

Poverty and Cultural Privation in the US

To gain any fundamental understanding of poverty studies in the US, there seem to be at least three significant areas of context that need to be grasped.

Official definitions

President Lyndon B. Johnson, in his 1964 statement of an all-out war on poverty, oversaw the development of a definition to mark the exact threshold of poverty and thereby to determine where to commit federal resources. The definition was also to serve as a marker against which to document the nation's march to victory.

The definition itself was based on the least generous Agriculture Department food budget for low-income families, and on assumptions that were made about household spending on other items (these assumptions remain largely unchanged). Any such definition is open to considerable criticism from all parts of the political spectrum. This description of poverty has been attacked from the political right because it does not include non-cash assistance or undeclared income in its calculations. More generally (although presumably from a more liberal political viewpoint) the definition has been criticized because, although it is adjusted for price inflation, it is not adjusted for rises in real income (Ruggles, 1990). As a result, the level of poverty in the US can be seen to fall even though there has been no redistribution of income and even though those at the bottom of the heap experience living standards that are increasingly below the social average.

Researchers in the US appear to be significantly constrained by having to use this official definition of poverty on which published government data are based. By other definitions, for example those households with less than half the national average income, poverty would be at a level twice as high as that recognized by official statistics. Even with the federal definition, there is little opportunity to take account of crucial local variations in food prices, and it would seem logical to expect that this also underemphasizes the plight of some fractions of the rural poor who have no easy access to a supermarket and therefore have to rely on 'mom and pop' stores where prices are higher and choice is restricted.

Economic restructuring

The second important area of context for poverty studies in the rural US is that of economic restructuring. In some senses this is a familiar pattern to European researchers. The switch from resources extraction and low-wage manufacturing industries (which have represented the backbone of most rural economies) towards service sector activities is, in general, similar to that occurring elsewhere in the developed world. Long-term decline in agriculture, forestry and mining has coincided with a rapid decline in mature manufacturing industry (see Chapter 1). The industries that had located in rural areas were low payers, very vulnerable to foreign competition, and offered very little protection to their work forces.

Although service sector growth has occurred in rural areas, it has tended to be located in the low-wage areas of consumer and personal services. Therefore, many rural areas in the US have suffered from an intensification of chronic economic instability, particularly where local industries have been volatile, unstable and vulnerable to cyclical trends.

Rural areas are thereby often dominated by low-wage employment in the agricultural manufacturing and service sectors, and where relatively high-wage employment exists it is often linked to 'boom-bust' industries, such as in mining and other resource extraction. These areas are also characterized by other jobs which are part-time or seasonal. Thus considerable underemployment exists leading to a significant incident of 'the working poor' (see below).

Problems arising from economic restructuring are not only related to a lack of economic growth or a lack of income. They are also associated with gross inequalities in the distribution of jobs and incomes within rural communities. Thus, black populations in the southern US, for example, are particularly vulnerable to underemployment, and rural women experience more limited employment opportunities than men and, therefore, experience higher poverty rates.

Two major responses to the problems of economic restructuring in rural areas have been noted by researchers in the US. First, there is continuing out-migration of impoverished households to urban centres of varying size where opportunities appear to be more plentiful. This serves as a rural drip-feed to apparently urban phenomena such as congregations of underclass population (see below). Second, there has been an increasing informalization of the labour market through the carrying out of 'odd jobs', the bartering of goods and services, and the rising levels of industrial homeworking.

The welfare response

The third area of context is that of the welfare response of poverty in the US. Welfare is delivered via a series of, to the European perception, rather restricted programmes, the main ones being: AFDC (Aid to Families with Dependent Children, particularly aimed at lone parents), Food Stamps, Medicare/Medicaid and Housing Assistance. Even given the restricted range of welfare programmes, Rucker (undated) has suggested that in 1985 some 65% of households with income below the official poverty line did not receive any programme assistance.

Welfare programmes have been subject to considerable political attack during the 1980s. Shapiro (1989) has described this as the federal government's 'direct policy to swell the ranks of the working poor'. For example, the 1981 Omnibus Budget Reconciliation Act transformed AFDC from a system that combined work with welfare to a system that encouraged a choice between work and welfare. The result was that 440 000 fewer households received AFDC 'at a stroke'. In addition, AFDC became linked with increasing levels of family break-up because of the stipulation that income will only be provided for a non-working woman without a husband.

It seems important to stress that programmes such as AFDC are in any case least effective in non-metropolitan areas. Three major factors contribute to this geographical bias:

1. The costs of the AFDC programme are shared between state and federal governments, but the level of benefit to be paid out is set by the state and there is no minimum level. Most of the non-metropolitan poor live in low benefit-level states. For example, in 1985 a one-parent family of three received US$550 per month in California and Vermont, but only US$96 per month in Mississippi.

2. The effectiveness of benefit programmes has to be set alongside the health care system in the US in which many of the rural poor are often uninsured or underinsured, and where rural areas are least well-served with facilities. A recent study by the Center for Disease Control (1990) has suggested that 'lack of income to pay for basic medical services is the most important factor in premature death in the US'.

3. There is often a very poor level of access to welfare in rural areas. Welfare offices are usually located in county seats, which can be very difficult and costly to reach for some rural residents. Moreover, there appear to be some cultural barriers for some poor rural families in dealing with 'the welfare folk'.

The Incidence of Rural Poverty

Against this background of economic restructuring, and of changing state responses to rural poverty, there have been a number of reports (for example, Barancik, 1990; Gorham and Harrison, 1990) documenting the scale of poverty, using government data (and the definitions inherent in those data). In 1989 there were found to be 9.7 million poor in non-metropolitan US, this being more than one in four of poor Americans and representing some 18% of the rural population of America. The number of rural poor appears to be growing.

A report by Porter (1989) has highlighted particular characteristics of the non-metropolitan poor:

1. 71.3% are white; 25.0% are black; 5.6% are hispanic (compared with 40.1% black and 23.8% hispanic in metropolitan areas);
2. 61.1% are in two-adult households and 38.9% are in one-adult house-

holds (compared with 4.17% in single adult households in metropolitan areas);

3. 12.6% are elderly (compared with a figure of 9.2% in metropolitan areas);

4. 64.6% have at least one paid worker in the household and 23.4% have two (compared with 51.1 and 15.9% respectively in metropolitan areas).

She concludes that the vast majority of the rural poor do not fit the common stereotype of poverty in the US, as most rural poor are white (indeed outside of the South nearly all the rural poor are white), poverty is primarily found in households headed by two parents and containing at least one paid worker, and the elderly are an important but minority subgroup of the rural poor.

There is a danger, however, that these aggregate data will themselves give rise to new and erroneous stereotypes. It is important to recognize many subgeographies of poverty lying beneath these aggregate figures with different rural localities being characterized by different social compositions, different patterns of economic restructuring, and different social relations working throughout the local state. For example Caudill (1983) stresses the importance of strong local elites in some states who will always want to run the state in the same way and who therefore represent very deeply rooted sets of social and power relations which will have enormous influence on the characteristics of localized poverty in the areas concerned. There are, therefore, many problems in explaining the specific existence of poverty in particular areas. How can the economic be linked with the cultural? How can we account for the feminization of poverty? To what degree is poverty self-sustaining through a lack of hope among generations of impoverished rural populations? Questions such as these necessarily refocus attention on the way in which rural poverty should be interpreted.

The Interpretation of Rural Poverty

If poverty is to be interpreted strictly as an economic condition, then researchers in the US have reached some consensus about its cause (Tickamyer and Duncan, 1990). First, poverty relates to economic restructuring, particularly where chronic economic instability and vulnerability have led to unemployment, underemployment and low pay. Second, and particularly in the South, poverty (for both blacks and whites) is rooted in a rigidly stratified political and economic system that perpetuates dependency and landlessness.

It does, however appear oversimplistic to conceive of poverty in purely economic terms, and certainly research on 'deprivation' must surely accommodate concepts that are wider than just economic poverty. Thus links are being made between economic poverty and life-style, and these links are portrayed in everyday semantics used to describe groups of low-income status as 'lower-class culture', 'low-income life-styles', 'culture of violence', 'slum centre', and even 'dregs culture'. It is at this point that the study and understanding of poverty can be subverted or sidetracked politically, and it is possible to trace analytical moments that represent change from recognizing particular life-style tracts associated with economic deprivation to blaming the undeserving poor for their own condition – for wallowing in a culture of poverty.

It was the anthropologist Lewis (1961, 1966a, b) who introduced the formal idea of a 'culture of poverty' in his ethnographic accounts of Mexicans and Puerto Ricans. He saw the 'culture of poverty' as a way of life that was passed down from generation to generation, and he suggested that such a culture was to be found in both urban and rural settings, resulting from a series of common adaptations to common problems. Moreover, he postulated that the 'culture of poverty' flourishes in economies where wages are low, unemployment is high, the social and political organization of the poor is underdeveloped, kinship is bilateral, and the dominant class values are associated with personal greed, personal gain and (on the flipside) personal inadequacy. As a result, the lower social strata in rapidly changing societies become alienated and marginalized. The 'culture of poverty', therefore, represents an adaptation to hopelessness and despair, because the people concerned realize that they are unable to achieve success in terms of the values and goals that are the norm in wider society.

Lewis saw the 'culture of poverty' as having several key features. For example, relating to family life: an absence of childhood as a specially prolonged and protected stage in the life-cycle, early initiation into sex, free unions or consensual marriages, relatively high incidence of abandonment of wives and children, and material dominance.

More generally he highlighted the key characteristics of: a weak ego structure, a lack of impulse control, a strong present-time orientation, and therefore little ability to defer gratification and to plan for the future, a sense of resignation and fatalism, and a high tolerance of psychological pathology of all sorts. Crucially, Lewis saw the surest way to eliminate the 'culture of poverty' as being through the organization of its members so that pride, organization and class or racial consciousness would lead swiftly away from any such culture.

Subsequently, it seems that the notion of the 'culture of poverty' has been taken up in America by both liberals and conservatives, but that it has been detached from the roots that were so carefully planted by Lewis. For example, Harrington's (1962) book, *The Other America*, was a key instrument in the rediscovery of poverty as an issue in the 1960s. In it, he defined the poor as those who for reasons beyond their control cannot help themselves. He stated that 'poverty is a culture . . . the family structure of the poor is different from the rest of society. There is a language of the poor, a psychology of the poor, a worldview of the poor'. Harrington sought, with some success, to arouse the conscience of the nation, but in so doing he lost sight of Lewis's focus on social organization. Rather, he saw intervention by sympathetic elites as the answer. Perhaps unwittingly, he thus permitted the 'culture of poverty' to become a political football to be kicked around the political field at will.

As political conservatism took greater hold in the US so, by the 1970s, the 'culture of poverty' had become a conservative concept:

1. the 'culture of poverty' resonated with traditional moral responses and definitions;
2. the 'culture of poverty' became seen as a self-perpetuated cycle – an excuse for prostitution, drug-abuse, crime and other forms of 'dysfunctionality' within the community;
3. therefore, the 'culture of poverty' became a justification for mean and punitive policies focused on the undeserving poor.

This appropriation of the concept of a 'culture of poverty' affected views on the definition of poverty, attitudes towards the unemployed, the role of public welfare, proclamations of the rights of the individual, and attributions for the responsibility for poverty. Such views are startlingly illustrated in Table 2.1.

The Underclass

By the 1980s the discourse on poverty had become dominated by the notion of an underclass. This conceptual switch is important for a number of reasons. It leads to a focus on the urban ghetto not rural poverty; on black one-parent poverty not male black or white poverty, or female poverty in two-adult households; and on dysfunctionality not the economic roots of poverty. In addition, it is a further illustration of how the poverty discourse has been requisitioned by conservative politics.

It again seems important to trace the underclass concept back to its roots. Although we should probably go much further back, it is perhaps sufficient in this context to mention the work of Gunnar Myrdal (1962) who reinvented the term 'underclass' to describe the chronically unemployed, underemployed and unemployable being created by postindustrial society. His focus was on those people who were being driven to the margins, or even out of the modern economy, and his concern was with reforming that economy not punishing the people who were its victims.

By the late 1970s in the US, the term 'underclass' had become identified

Table 2.1. Conservative appropriations of the concept of poverty.

The Definition of Poverty
We must avoid any definition that implies the need of a level of help . . . that would tempt the recipient to become permanently dependent on it, and undermine his incentives to self-support. (p.38)

The Unemployed
These include the aged, weak, disabled or blind. They include the feeble-minded, and the people so backward that they cannot be taught elementary skills, and require more supervision than it is practicable to provide. They include, finally, the chronic loafers – the Rip Van Winkles, born with an insuperable aversion to all kinds of profitable labour – and the hostile, who refuse to accept any kind of discipline. (p.108)

The Role of Welfare
An able-bodied adult needs a level of subsistence sufficient to maintain reasonable health and strength . . . Any relief program that tries to provide more than this for idle able-bodied adults will in the end do more harm than good. (p.39)

The Rights Of The Individual
There is one political change that is practically imperative if a nation is not to be driven towards bankruptcy by relief and redistribution programs completely out of control. This is to suspend the right-to-vote of anybody on public relief. (p.202)

The Responsibility For Poverty
Basically, each individual – or at least each family – must solve its own problem of poverty. (p.230)

Source: compiled from Hazlett (1986).

with acute and persistent poverty (rather than joblessness) and had thus been attributed a new definition comprising three significant elements:

1. a racial element: the underclass is thought to be black and/or hispanic;
2. a behavioural element: the underclass is characterized by behavioural patterns which conflict with mainstream values – joblessness, out-of-wedlock births, welfare dependency, school dropout, and illicit activity;
3. a geographical element: the underclass is thought to reside in segregated deprived neighbourhoods where social problems are highly concentrated, resulting in mutually reinforcing contagious effects through imitative behaviour and peer pressure.

These new definitions have been reinforced by some of the more recent journalistic treatises on underclass (for example, Auletta, 1982; Lemann, 1986, 1988). They have also been reinforced by the attempts of academic researchers to measure the American underclass; these have been reduced to the definition of urban neighbourhoods with unusually high amounts of 'antisocial' and 'deviant' behaviour, as if such neighbourhoods were demarcated by such behaviour (see for example, Ricketts and Sawhill, 1988; Hughes, 1989; Ricketts and Mincy, 1990). These analytical developments have led to the term 'underclass' taking on connotations of undeservingness and blameworthiness, and to the polluting of poverty discourse with ideological overtone and political implication from the conservative right.

Some re-evaluation of the concept of underclass has been inspired by the work of the Chicago sociologist William Wilson, especially in his book *The Truly Disadvantaged* (1987). In an earlier thesis (1976), Wilson stressed the emergence of class stratification among blacks. A black middle class had moved into better jobs and better neighbourhoods, but other groups of blacks had been left behind in the inner cities in a deepening crisis; Wilson regarded these as the underclass. In *The Truly Disadvantaged*, Wilson argues that the notions of 'working class' or 'lower class' are unable to explain the spatial concentration of poverty in inner cities and the harmful effects of the isolation of black groups there from other parts of the black community. Instead he points to a distinct territory of social concentration, a social existence outside the mainstream US occupational structure, and the development of behaviour at variance with mainstream patterns and norms, and suggests that these factors are unprecedented historically and, therefore, represent a new facet of class formation.

Even Wilson's work has been appropriated by conservative ideology and by the efforts of those who are determined to provide a quantitative measurement of the underclass. Wilson's subjects have thus been interpreted as the epitome of the 'dregs' culture associated with the undeserving poor and the culture of poverty, and his complex analysis has been subverted to suggest a measurable underclass which consists of urban, black, female-headed households with a detachment from formal work and a high incidence of behaviour which is deviant from social norms.

A Rural Underclass?

It is important to stress the way in which different meanings have been ascribed to concepts such as 'culture of poverty' and 'underclass', since these meanings set the agenda for interpreting poverty more generally. What is evident from the above is that rural areas in America are not seen as locales inhabited by fractions of the underclass and, therefore, by implication they are not locales where the most important and qualitatively 'worst' forms of poverty exist. As in Britain, rural poverty has become out of sight and out of mind. Rural poverty in America is associated with two-adult households, the working poor, a lack of social isolation, low levels of dysfunctionality, and now welfare dependency – the antithesis of the defining characteristics of the underclass.

Nevertheless, it can be argued that, apart from the spatial concentration which is evident in inner-city ghettos, most of the defining attributes of the underclass are relevant to rural poverty in America. At this point the paucity of this author's own knowledge becomes very apparent, but it seems possible to suggest three strands of evidence with which to support the previous assertion. First, there is a very real sense in which the problems of living in rural areas are part of the feeding of urban inner-city problems through rural-to-urban migratory movements. It would, therefore, appear rather dangerous to permit what seems like a distinctive geographical difference between 'rural' people and 'urban' people to be translated necessarily into distinct social differences.

Second, there are clear examples of a concentrated form of underclass in rural areas. Research work focusing on Indian reservations (for example, Sandefur, 1988, 1989) has shown that if well-established measurements of underclass are used (for example, poverty, welfare dependency, social disorganization etc.) many reservations score as highly as do inner city ghetto neighbourhoods. Here, surely, is a clear case where economic, social and physical isolation from the majority of society has resulted in extreme poverty, high unemployment, unstable families, low rates of high-school graduation, and high rates of alcoholism, drug use and crime.

Third, it can be further argued that even where rural poverty is not geographically concentrated, as in the case of Indian reservations, many of the characteristics of an underclass exist even though they may be more hidden and take on a slightly different form. Using admittedly often anecdotal evidence from the extensive literature on Appalachia, it is possible to question the major assumptions that deny the existence of a rural underclass.

Fewer lone-parent households?

This assertion has some statistical validity, but there is extensive household disruption, instability and vulnerability associated with rural poverty. Do the cultural traditions of early marriage and a reluctance to divorce actually make much difference to the poverty experienced in rural areas such as Appalachia? Such traditions appear in any case to be slowly disintegrating. Is rural poverty really any less feminized than its urban counterpart? Emerging ethnographic narratives of rural poverty suggest not.

Less detachment from the labour force: more working poor?

According to Porter's (1989) analysis (quoted above) there is only a 13.5% difference between the numbers of working poor in non-metropolitan and metropolitan areas, although it might be considered that the 'working poor' and the 'non-working poor' should be considered as different fractions of an impoverished society. Rural poverty is not solely connected with the working poor; however, Tyrel Moore (personal communication) suggests that, on the basis of his extensive research in Appalachia, it is possible to demonstrate that there are many poor households who would not work even if they had the opportunity. Moreover, there are people in menial work who take no pride whatsoever in that work and cannot, therefore, be described as fully attached to the labour force. In addition, to some extent the inaccessibility of, and low levels of payment from, welfare programmes such as AFDC mean that an urban-type welfare dependency is not possible in many rural areas. In these situations, the informal and semiformal economy has to be exploited simply because there is no welfare safety net.

No social isolation?

The work of Tickamyer and Tickamyer (1987) and others shows that socioeconomic and cultural isolation have been the experiences of many generations of rural poor, especially in the South where rigid social stratification has kept them out of the mainstream.

No dysfunctionality?

It is also possible to refute this suggestion. There appears to be evidence of widespread alcoholism in rural areas, and of associated problems in poor households such as the battering of women and children. Equally, dysfunctionalities relating to drug use and abuse cannot be confined to urban inner-city areas. Drug production and use are becoming increasingly important in rural America; marijuana is said to be Kentucky's leading cash crop over the last 5 years or so. In one sense drugs have become part of the necessary informal economy, given the problems relating to the restructuring of the formal economy. Such production does, however, ensure that drugs are freely available for local use, and so there seems to be little evidence of crime linked to the need to obtain drug money. This does not suggest an absence of criminality, however, and high rates of crime in many rural areas are borne of the general sense of hopelessness and frustration and of the resentment against the conspicuous consumption of possessions by wealthy rural residents, and especially by wealthy tourists.

These arguments require a great deal more thought and research than is presented here, but these indications do suggest the existence of underclass phenomena in rural America. If we return to Myrdal's notion of underclass, there appears to be a very clear presence of impoverished class fractions in rural areas. Further, if we accept the urban-based concepts of underclass of Auletta, Lemann and even Wilson, it can be suggested that similar attributes exist in rural areas, even though some characteristics are differently constituted. Such assertions do seem to open up important implications for the theorization, and therefore greater understanding, of the changing nature of rural poverty.

Pointers to Research in Britain

There are at least four implications for research in Britain from this tentative interpretation of the study of poverty in America. None of the four are innovative – indeed similar conclusions could probably have been reached using different terminologies from a study of the historical geography of poverty in Britain in the 19th century. Nevertheless, the experience from the US underlines the importance of a conceptual as well as an empirical understanding of poverty and the cultures and life-styles that surround it. The four implications are outlined below.

The incidence and measurement of poverty

The existence of an 'official' measure of the poverty line is a mixed blessing. It has permitted extensive empirical analysis of poverty at an aggregate level, and at least government cannot claim that poverty does not exist when their own (however defective) indicator says otherwise. On the other hand, the existence of a poverty line allows the poor to be herded to one side and dismissed as 'irresponsible' and 'undeserving'. In some ways the emphasis on defining and measuring poverty in the US seems to have aided the political appropriation of the concept of poverty. However, much of the official definition is criticized; it has always been easier to use it than not, especially given the huge difficulties in gathering non-governmental data for any extensive area of the US.

In the UK there is no such definition, and therefore governments can (and do) suggest that poverty does not exist. A recent book by Oppenheim (1990), suggests that 10% of Britain's population is living in poverty (defined as less than 50% of the average wage), but again such analyses can easily be disputed by those in power. The truth is that we do not know the extent of rural poverty in Britain, and even if we did have an official definition, there seems little doubt that the figure would be regularly 'massaged' downwards (as has happened, for example, in the frequent redefinition of 'unemployment' in Britain). This serves to emphasize the need for more empirical research to ascertain the extent of and changing nature of rural poverty in Britain.

The appropriation of concepts of poverty

In the UK we are experiencing the same political appropriation of concepts relating to poverty as is evident in the US. For example, the idea of blaming the poor for their own culture of poverty has been rife in Thatcher's Britain. To quote from Ferdinand Mount, adviser to 10 Downing Street in 1982:

> The rich admire the poor less and less, partly because the poor are not as poor as they used to be, but also because the poor fritter their money on trash – video cassettes and cars with fluffy dice that joggle in the back window.
>
> (quoted in Walker and Walker, 1987, p. 11)

The need for a welfare response to poverty has been equally sanitized. Government minister, John Moore, said in 1987:

> we believe the well-being of individuals is best protected and promoted when they are helped to be independent, to use their talents to take care of themselves and their families, and to achieve things on their own.

Most notably, the concept of the underclass has become so polluted that even pressure groups such as the Child Poverty Action Group now want to disclaim it:

> we have argued that the concept of an underclass responsible for its poverty by virtue of its own behaviour cannot be sustained.
>
> (Oppenheim, 1990, p. 16)

The reclaiming of concepts

Rather than abandoning important conceptual tools of analysis that have been polluted ideologically, it is argued here that we need to reclaim what has been misappropriated. This is important in two respects.

Underclass and class fractioning

Class-based analysis of social recomposition in rural areas must be able to identify common structural positions within the social organization of production, based on ownership or non-ownership of factors of production. Underclass will, therefore, be characterized by non-ownership of saleable labour power. We need a much fuller analysis of the relationship between the underclass and 'working' and 'capitalist' classes that will, at the very least, involve a discussion of the role of underclass as a reserve army of labour. We also need a much clearer idea of how poverty is linked with different class fractions, and how the wider condition of deprivation is similarly linked with different class fractions. This will in turn aid our further understanding of inter- and intraclass conflict within wider social recomposition.

Cultures associated with poverty

It seems fundamental to emphasize the life-styles that are associated with poverty, impoverishment, social and economic marginality and wider aspects of deprivation. However, such emphasis must be dissociated from 'blaming the victim'. In turn, this identification of 'cultures' and 'life-styles' may help us to unravel the current problems of defining and focusing concepts such as deprivation and disadvantage. Accordingly, we might attempt to identify:

1. social fractions in poverty because of underclass location, or other class locations associated with underemployment and the working poor;
2. social fractions that are culturally impoverished through factors such as a lack of security, access, opportunity, choice or power. Matters of gender, age, ethnicity and handicap will all be crucial in this respect.

The need for a conceptual framework

As argued elsewhere (Cloke and Goodwin, 1991), it is crucial to match the broad processes discussed here with particular people in particular places. There is therefore a need to bring together economic restructuring, social recomposition and the role of the state in a theoretical context that explains difference as well as similarity. One potentially fruitful framework is that of recognizing the structuring of coherence within a post-Fordist ensemble of relations, and one of the aims of the current study of life-styles in rural Britain is to place new empirical material into this conceptual context.

Acknowledgement

I acknowledge with grateful thanks the support of the Arkleton Trust in undertaking the work on which this chapter is based.

References

Andrews, K. and Jacobs, J. (1990) *Punishing The Poor: Poverty Under Thatcher.* Macmillan, London.

Atkinson, A. (1989) *Poverty And Social Security*. Wheatsheaf, Brighton.

Auletta, K. (1982) *The Underclass*. Random House, New York.

Barancik, S. (1990) *The Rural Disadvantage: Growing Disparities Between Rural and Urban Studies*, Center on Budget and Policy Priorities, Washington DC.

Barooah, V., McGregor, P. and McKee, P. (1991) *Regional Income Inequality And Poverty In The United Kingdom*. Dartmouth, Aldershot.

Brown, P. and Scase, R. (eds) (1991) *Poor Work: Disadvantage And The Division of Labour*. Open University Press, Buckingham.

Caudill, H. (1983) *Theirs Be The Power*. University of Illinois Press, Chicago.

Center for Disease Control (1990) quoted in *Atlanta Journal and Constitution* 30, September, p. A1/A10.

Cloke, P. (ed.) (1992) *Policy And Change In Thatcher's Britain: A Critical Perspective*. Pergamon Press, Oxford.

Cloke, P. and Goodwin, M. (1991) Conceptualising countryside change: from post-Fordism to rural structural coherence. Unpublished paper presented to the annual conference of the AAG, Miami, Florida.

Glendinning, C. and Millar, J. (eds) (1987) *Women And Poverty in Britain*. Wheatsheaf, Brighton.

Gorham, L. and Harrison, B. (1990) *Working Below The Poverty Line*. Department of Urban Studies and Planning, Massachusetts Institute of Technology.

Harrington, M. (1962) *The Other America: Poverty In The United States*. Macmillan, New York.

Hazlett, H. (1986) *The Conquest of Poverty*. University Press of America, New York.

Hughes, M. (1989) Misspeaking truth to power: a geographical perspective on the underclass fallacy. *Economic Geography* 65, 187–207.

Lemann, N. (1986) The origins of the underclass, *The Atlantic Monthly* July, 54–68.

Lemann, N. (1988) The unfinished war. *The Atlantic Monthly* December, 37–56.

Lewis, O. (1961) *The Children Of Sanchez*. Random House, New York.

Lewis, O. (1966a) *La Vida: A Puerto Rican Family In The Culture Of Poverty*. Random House, New York.

Lewis, O. (1966b) The culture of poverty, *Scientific American* 215, 19–25.

Lowe, P., Bradley, T. and Wright, S. (eds) (1986) *Deprivation And Welfare In Rural Areas*. Geobooks, Norwich.

McLaughlin, B. (1986) The rhetoric and reality of rural deprivation. *Journal of Rural Studies* 2, 291–307.

Myrdal, G. (1962) *The Challenge To Affluence*. Pantheon, New York.

Oppenheim, C. (1990) *Poverty: The Facts*. Child Poverty Action Group, London.

Porter, K. (1989) *Poverty In Rural America: A National Overview*. Center on Budget and Policy Priorities, Washington DC.

Ricketts, E. and Mincy, R. (1990) Growth of the underclass, 1970–80. *Journal of Human Resources* 25, 137–45.

Ricketts, E. and Sawhill, I. (1988) Defining and measuring the underclass. *Journal of Policy Analysis and Management* 7, 316–25.

Rucker, G. (undated) *Rural Poverty In Perspective*. Rural Coalition, Washington DC.

Ruggles, P. (1990) *Drawing The Line: Alternative Poverty Measures And Their Implications For Public Policy*. Urban Institute Press, Washington DC.

Sandefur, G. (1988) Blacks, Hispanics, American Indians, and poverty – and what worked. In: Harris, F. and Wilkins, R. (eds) *Quiet Riots: And Poverty In The US*. Pantheon, New York, pp. 46–74.

Sandefur, G. (1989) American Indian reservations: the first underclass areas? *Focus* 12, 37–41.

Shapiro, I. (1989) *Laboring For Less: Working But Poor In Rural America*. Center on Budget and Policy Priorities, Washington DC.

Tickamyer, A. and Duncan, C. (1990) Poverty and opportunity structure in rural America. *Annual Review of Sociology* 16, 67–86.

Tickamyer, A. and Tickamyer, C. (1987) *Poverty In Appalachia*. Appalachian Center, University of Kentucky, Lexington.

Walker, A. and Walker, C. (eds) (1987) *The Growing Divide: A Social Audit, 1979–87*. Child Poverty Action Group, London.

Wilson, W. (1976) *Power, Racism and Privilege: Race Relations In Theoretical and Sociohistorical Perspectives*. Free Press, New York.

Wilson, W. (1987) *The Truly Disadvantaged: The Inner City, The Underclass, And Public Policy*. University of Chicago Press, Chicago.

ASPECTS OF SOCIAL CHANGE IN RURAL AREAS

3

THE TRANSFORMATION OF RURAL LIFE: THE CASE OF TORONTO'S COUNTRYSIDE

Michael Bunce and Gerald Walker

Toronto's Countryside and Canadian Exurban Development

Cities in Canada have experienced the same suburbanization and exurbanization processes as in the United States and elsewhere in the capitalist world. This has been accompanied by an interesting cultural appraisal of the evolving peri-urban environment. The popular view is that the cities are a constant threat to rural life in general, and to agriculture as the base of rural life in particular. This view is widely shared by urbanites and ruralites alike. Yet, the relatively little research carried out to date has indicated that the actual transformation of rural, again largely agricultural, land uses into urban has been marginal. Yeates (1975) found, in his study of the Windsor to Quebec corridor, that the absorption of rural land into urban uses was not significant. His findings raise questions about the relationship of cities to their countrysides.

It is contended here that the changes that have transformed the countryside were mediated by the property system and the land market, including both the traditional agrarian landed interests and the newer exurban landed interests. Two critical elements are developed in this argument to explain the transformations: endogenous agricultural restructuring, identified with the ruralization process; and exogenous capital investment, based on the commoditization of amenity.

The movement of urbanites into the Canadian countryside, as in most other western industrialized countries, began in earnest during the 1950s. By the early 1970s rapid and largely uncontrolled expansion of exurbanite residential development had occurred around most major cities. The only study at the national scale is that by Bryant and Russwurm (1984). Their analysis of 52 urban fields revealed a complete reversal in the ratio of farm to country residential population between 1951 and 1976, with the most rapid growth in the latter occurring after 1973 (Fig. 3.1). Although

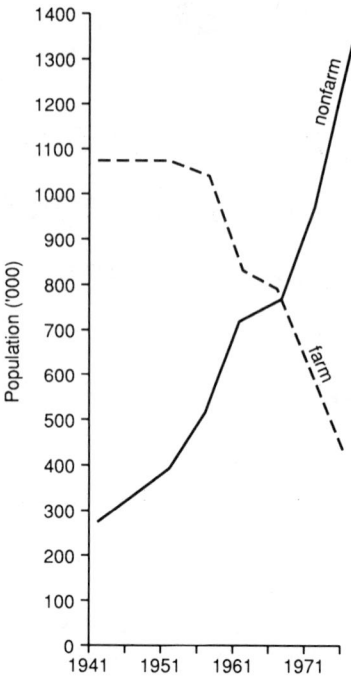

Fig. 3.1. Farm and scattered non-farm population in Canadian regional cities.
Source: Bryant and Russwurm (1984, p. 115).

there was considerable regional variation in this trend, it is clear that it affected all the metropolitan regions of the country. This is confirmed by local case studies (McCrae, 1980), most notably of the Prairies (Momsen, 1984) and of the Montréal-Québec region (Lepine and Brunet, 1984). In the Edmonton-Calgary corridor, in particular, there was a veritable boom in the creation of residential lots in rural areas from the late 1960s.

Toronto has been the most dynamic of Canadian urban regions. In the early 1970s, the urban zone as defined by Statistics Canada (1974, 1977, 1980, 1988a, b, c) expanded dramatically. This expansion continued into the 1980s. Much of the peri-urban zone of Toronto effectively became urbanized between 1970 and 1990. The population transformation and urban expansion has restructured all places in the Toronto countryside. Much of the fundamental social geography, as well as the economic relationships have been transformed. The settlement transformation may be summarized by looking at typical places in Toronto's countryside.

There are four kinds of places in the Toronto countryside: exurbs, suburbs, rural segments and small towns. Exurbs – very low density urban settlements – dominate much of the urban shadow (Fig. 3.2). Places such as the Caledon Hills, the Albion Hills, Wildfield, King Township,

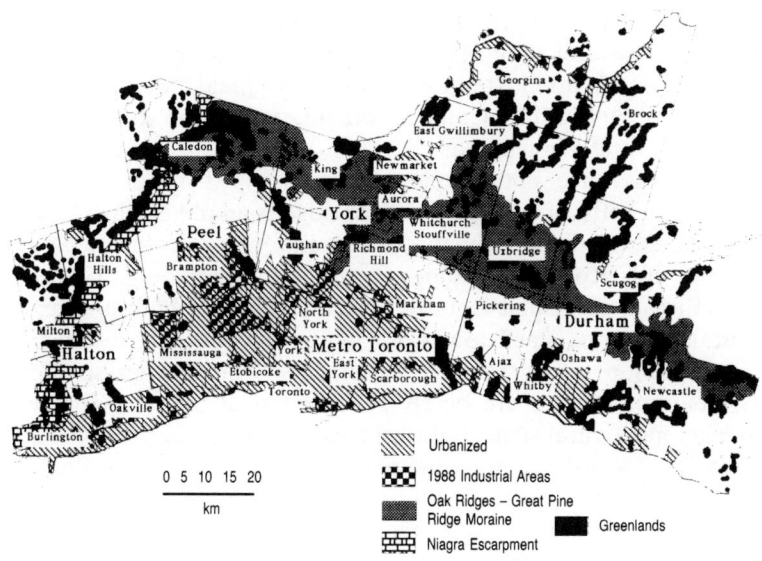

Fig. 3.2. Greater Toronto area. Source: Kantor (1990), Greater Toronto Coordinating Committee (1990).

Whitchurch, Kleinburg and rural eastern Vaughan are all dotted with small subneighbourhoods, with highly localized interaction fields. These are articulated into larger zones, in part through the zoning and planning designations of local municipalities. These subneighbourhoods are frequently anchored on scattered estate developments and ribbon settlements, and have a core of dispersed urban dwellings in the midst of an agrarian or wooded setting. In the Toronto setting, exurbs coexist with rural segments (Walker, 1987).

Suburbs – low-density nucleations – are the typical settlements for the fringe, particularly the inner fringe, of the urban field. Markham, Thornhill, Maple, Woodbridge, Nobleton, King City, Heart Lake, and the fringes of Aurora and Newmarket are all suburban in structure. Small to middle-sized neighbourhoods are the most frequent interactional fields. Very commonly, one of the several neighbourhoods in a suburb will be dominant and will have its interaction field interpenetrate hierarchically with the remaining neighbourhoods. Suburbs typically encircle old, small towns.

Rurality does survive in the Toronto countryside, though in overlap with exurbs and suburbs. Rural segments – extended farm-centred neighbourhoods – survive in most of the exurbanized countryside. Large, usually kinship-based neighbourhoods, which are highly mobilized, are found in the Caledon and Albion Hills. The Peel Plain, south of Caledon, is divided into several rural neighbourhoods, as is King Township. Rural segments survive in attenuated form around Wildfield and in the undeveloped portions of Brampton and Markham, as well as in the Town of Vaughan and rural Whitchurch.

Small towns have not disappeared from the Toronto countryside. Buried in the centre of Italianate Woodbridge is an old English Canadian village. Nobleton still has remnants of a small town. Aurora and Newmarket, along with Holland Landing, Stouffville, Schomberg, Bolton and Caledon East are all small Ontario towns in their old built-up areas. As with rural segments, small towns are in varying states of disintegration and reintegration into urbanized settlements. The first group of small towns mentioned above are barely small towns at present. The second group have more rural structural elements, and it is still unclear whether these small towns will integrate the new suburbs or vice versa. Places like Malton and Brampton, though still containing elements of small town life, are essentially urban in character today.

It is low-density, exurban, residential development that has dominated the region as a whole, and where, as the study by Bryant and Russwurm (1984) shows, exurbanization trends first appear. From a plethora of case studies, there emerges the now familiar picture of an Ontario countryside during the 1960s and 1970s affected by a rapid and widespread growth in scattered residential development. Studies have documented the proliferation of country residences around London and Kitchener (Russwurm, 1976; Troughton, 1976) as well around smaller centres, such as Guelph (Joseph and Smit, 1985). Particular attention has been paid to the Toronto region. A field survey, carried out between 1969 and 1972, revealed a dense pattern of 'rural estates' in an arc extending up to 50 km from the Metropolitan Toronto boundary (Michie and Found, 1976) (Fig. 3.3). Much of this, as Punter's outstanding study has shown, was facilitated by the boom in severance activity (the severing of small land parcels from larger, usually farm holdings) in the Toronto fringe during the 1950s and 1960s which created literally thousands of rural lots in the 1–10 ha range (Punter, 1974) and a ready supply of land for those seeking a residential retreat in the countryside.

By the late 1960s, the level of severance activity and conversion of severed parcels into country residences had reached such proportions that municipal and provincial policies were introduced to control the process. Although this was effective in significantly reducing the number of severances and in diverting residential development away from a dispersed pattern, especially in the most pressured municipalities, to a large extent the die had been cast. By the early 1970s exurbanites already owned between 25 and 35% of the land in Punter's study areas. And, by imposing lower limits on the size of severances as a means of limiting their numbers,

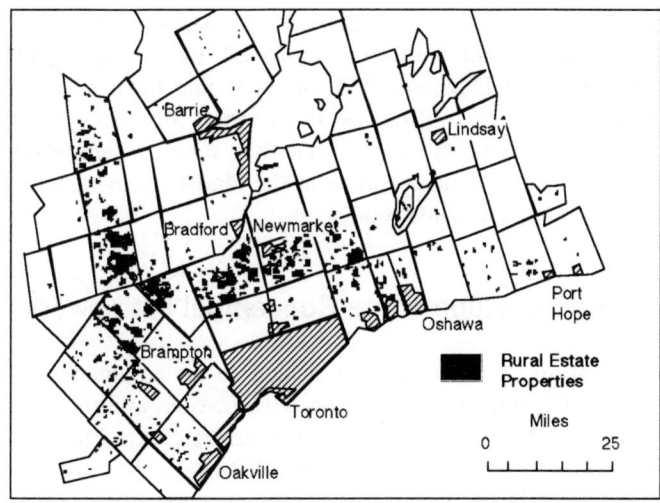

Fig. 3.3. Rural estate properties in the Toronto Region, 1968–1972. Source: Michie and Found (1976, p. 18).

municipalities did little to reduce overall land supply for country residences. Moreover, even after controls had been introduced, most municipalities still contained large numbers of undeveloped land parcels for which severance consent had already been granted. Finally, in areas of marginal value for agriculture, the purchase of whole farms by exurbanites could continue without the need for severance approval.

Throughout the 1970s, therefore, the earlier boom in severance activity translated into a continued, if somewhat slower-paced process of country estate development. As the supply of prior consents dried up and the introduction of provincial farmland preservation guidelines forced municipalities to restrict further the granting of new ones, the rate of this development in some municipalities slowed to a trickle, especially those in the inner fringe where urbanization pressures were the greatest. Yet, as Punter's study has shown, and as Walker's (1987) work has confirmed more recently, the country residences of exurbanites had already become a dominant element of the landscape and society of the Toronto region. Furthermore, there is some evidence that exurbanization pressures have re-emerged in the past few years. One of the reasons for this is that, in restricting severances, municipalities have increasingly directed new residential development in the countryside into rural subdivisions – clusters of houses on unserviced lots of generally between 0.5 and 1.5 ha – designed to provide a measure of private amenity without the sprawl of individual estates. Data from the Regional Municipalities of Durham and

York show that, over the last 15–20 years, almost 4000 housing units of this kind of development had been approved and registered (Durham, 1991; York, 1991). In Durham, 75% of these had been registered since 1980. Although there are no figures on the total area occupied by these subdivisions, their large lots suggest that they take up significant chunks of rural land. Certainly their presence in the landscape is all too apparent as one drives around the countryside.

Fringe Development: a Conceptual Framework

The transformation of rural districts is due to an amalgam of causes – those exogenously centred in the city or urbanization, and the endogenous forces of economic reorganization and structuring centred in the country-side, both ultimately deriving from contradictions in the capitalist mode of production. Popularly, urbanization influences have been seen as most significant, although Yeates (1975) found contradictory evidence. A recent study found that farmers who had relocated from the rural-urban fringe explained their decision almost entirely in terms of the low returns to agriculture, not the impact of urban forces (Maitland, 1988).

These juxtapositions suggest that the rural environment is not a fragile shell just waiting for the impact of urban invasion. Rural life is commonly threatened, without regard to how near the large city is to the particular rural segment. At the same time, rural life is very resilient. The train of events seems to suggest that urbanity is the weaker element in the transformation. As the countryside empties, it presents opportunities for urban encroachment. Speculation and the transformation of the land market are significant aspects of the urbanization of the countryside, but not because rural life cannot or does not resist. Urbanization is a weak force, a force that moves into a near vacuum created by the stronger force of ruralization. Clout (1972) identified this force in the early 1970s with three constructs visible in that period: depopulation, occupational simpli-fication and demographic restructuring. These aspects of the endogenous character of rural transformation, quite independently of the forces of urbanization, still seem to be salient.

The underlying cause of ruralization is in the poor, and declining, returns to agricultural activity. Farmers, as a pivotal class in the rural environment, simply cannot make a living. Others, depending on the circulation of incomes and capital from the farmers, become vulnerable and disappear from the rural scene. The effective centre of the web of causes that produces ruralization is capitalism and its continued transform-ations of all environments within which it is situated. Capitalism is, of course, just as dominant in the city as in the countryside. The city is the great heat pump of the capitalist system. The capital realized in the city, and attracted or siphoned to the city, is constantly in search of opportunities for investment everywhere. The core of the weak force of urbanization is the necessity for reinvestment of capital. Speculative land markets are one of the ways this process of capital movement is acted out. The cultural and social dominance of the city is the weak periphery of urbanization, not its underlying cause.

While rural segmentation produces more or less recognizable collective structures, reinforcing and recreating a culture, it is highly privatized and class divided. Property, and particularly landed property, is the base of rural community. The central struggle is to maintain, perhaps even to augment, propertied position. The surviving farmers are the most archetypical representatives of this trait. There are other rural residents – descendants of labourers who now work in nearby towns, former farm families who still hold but do not farm the land and work in town, rural professionals and rural merchants – all of whom look to the property system to define them and to provide the tangible base for their social positions.

The farmers' reorganization includes a sharpening difference between substantial full-time farmers, with high productivity (albeit with high degrees of economic vulnerability), and part-time farmers, who have largely become asset holders rather than producers (Agricultural Council of Ontario, 1986). Urban expansion into that countryside in which farmers were declining has meant a social and economic invigoration of the rural areas. The process continues, with both groups responding to the reorganization of capitalism by generating new social spaces.

Urbanites appear to pay much less attention, at least overtly, to property, but it is no less important to them. Urbanites do not try to accumulate land as do the rural population. Nevertheless, the urbanites, mainly originating from the privileged classes of Toronto, recognize that the purchase of a home is a critical social statement as well as a necessary investment. In most of the countryside of Toronto there is a complementarity between the property interests of the rural and urban segments. Urbanites have quite modest property acquisition aspirations. When forced to buy more land than they have use for, they are very happy to rent the land to local farmers. Furthermore, most of the time the existence of farmers is considered an aesthetic advantage to the urbanites. Occasionally conflicts surface, as when farmers spread manure and pesticides upwind from urban residents. Usually the two segments have a tacit understanding that allows joint occupancy. The secular rise of land values is seen as advantageous to both segments; it is the farmer's insurance policy and the urbanite's guarantee of privilege into the next generation.

The property system, as fundamental as it is to capitalism, is only one of the preconditions for capitalist exploitation. The capitalist transformations of the last generation – away from fordist labour/capital concentrations and toward flexible production – have had their effects on Toronto's countryside. Canada's agriculture has moved dramatically away from a *genre de vivre* to become an industry. Central to that transformation has been the reorganization of the farmers' class (Walker, 1988). The crises of ruralization are another way of expressing the effects of the long-term transition from petty commodity production to industrial production in the countryside (Volume 1 develops this theme).

Exurbanization is also part of the metropolitan decentralization process and is thus to some extent driven by general externalities such as the housing and job markets. Yet, the acquisition of a country property remains primarily a search for personal amenity. That it reflects an attraction to the stereotypical and somewhat mythological ambiance of the countryside – open space, natural surroundings, tranquillity, country life – has been well documented, both generally and in the specific context of southern Ontario (Beesley and Walker, 1990; Bunce, 1981; Davies and Yeates, 1990). The main concentrations of exurbanites in the Toronto region are certainly in the more scenically attractive and secluded areas of the Niagara Escarpment and the Oak Ridges Moraine. Equally important, however, are the amenities of the country property itself. Indeed the qualities of the external setting are often of secondary importance to those of the internal residential landscape. A few hectares of land provide not only space and privacy, but also the opportunity for the expression of personal amenity aspiration, which Spectorsky has argued is the underlying motivation of exurbanites (Spectorsky, 1955). This covers a wide range of pursuits, from hobby farming to casual enjoyment of well-landscaped grounds (Punter, 1974). While this reflects the individualism inherent in the demand for a country retreat, it can also be influenced by ethnic and class differences (Walker, 1987).

The conversion of large areas of the countryside into a residential landscape is at heart driven by the demand for ownership of amenity resources. With ownership comes control over the use of these resources, the assurance of unrestricted and exclusive access to their enjoyment, and the freedom to fashion personal amenity spaces. As already suggested, this is manifested in a variety of ways in which exurbanites express the amenity value of their property. In the Toronto region this has produced a fairly eclectic exurban landscape that reflects individual perceptions and expectations of the countryside (Punter, 1974). Yet, underlying this diversity is the common denominator of the maximization of the use value of amenity that derives from private ownership. This imputes a new kind of value to rural land in which productive uses are replaced by the consumption of the more intangible qualities of the environment.

The privatization of countryside amenity, however, is not limited to its use value. Land, of course, has exchange value, which is measured in terms of both its current and potential uses. In the pressured land market of an expanding metropolitan region, it is the development potential of land that determines its value. We can thus see exurbanization as a process in which the amenity attributes of rural land acquire exchange value in their own right. This can be conceptualized in terms of a commoditization process. Briefly explained, commoditization involves the extension of the commodity form to new areas of activity or the formation of new types of commodity relations (Marsden, 1989). In the case of country residential development, new use values centred on amenity attributes are created, which are then, in turn, commoditized into exchange value (Marsden and Murdoch, 1990). That this characterizes the exurbanization process in the Toronto area is evident in the general inflation of the exurban land market, and in particular in the recurrent speculative activity in land severances and subdivisions. The tightening of severance controls has put further pressure on the market for country estate property. It has also heralded the growth of developer activity in rural subdivisions. Indeed, nowhere is the commoditization of the amenity value of the countryside more explicit than in the marketing of these developments (Bunce, 1981).

Central to the idea of commoditization, so far as the restructuring of rural areas is concerned, is the importance of external capital to the reorganization of land and property relations (Marsden and Murdoch, 1990). This can be readily applied to the exurbanization process, for clearly the transformation of rural land into amenity property depends largely on the injection of urban-derived income into the land market, as well as into the development and improvement of property itself. The argument here is that the sustenance of this landscape depends upon the reproduction of landed capital, rather than on simply the use value of amenity. In other words, the ultimate worth of a country property and, therefore, the justification for maintaining it, lies in its investment and exchange value. We can thus see the commoditization of amenity as the process by which exurbanites, and those who control the exurban land market, restructure both the physical and economic rural landscape around their own financial agenda. Viewed from the political economy perspective of commoditization theory, this occurs through the re-ordering of local power relations, in which the owners of new forms of landed capital intrude upon and eventually displace more traditional landed interests.

This thesis, of course, is consistent with the class analysis that has dominated the discussion of exurbanization for years, which argues that urbanites moving into rural areas constitute a new elite which comes to dominate the local political agenda. Although several scholars have shown that class relations in exurbia are considerably more complex than this (Ambrose, 1974; Walker, 1987), there is strong evidence that exurbanites often control local amenity movements (Lowe, 1977), and are thus in a position to direct local affairs around their own property interests.

Furthermore, the commoditization of rural amenity extends beyond the boundaries of exurbanite property and into the rural economy in general, through the refashioning of exurban places around the culture of country-style consumerism (Coppack, 1988). In rustic inns and country restaurants, antique shops and craft boutiques, village fairs and summer music festivals, the culture of exurbia has become a profoundly commoditized experience. It is woven seamlessly into a preservationist ethic in which the creation of rural authenticity goes hand in hand with commercial opportunity. In a recent study of an exurban community in Pennsylvania, Dorst (1989) argues that this is representative of the culture of advanced consumer capitalism, of the culture of post-modernity reflected in the transformation into a general commodity form, which encompasses everything from residential property to the character of the rural landscape as a whole.

Conclusions

This chapter has sketched the development of the urbanization of the countryside in Canada, with emphasis on the contexts of central Canada and, in particular, the Toronto region. One element that has linked together all developments is the mediation of change by the property system; that property system encompasses both traditional agrarian landed interests and newer exurban landed interests.

Two elements of interpretation have been suggested: endogenous agricultural restructuring, identified with the ruralization process; and exogenous capital investment, based on the commoditization of amenity. These two aspects of the restructuring of landed capital are also linked to the social characteristics of the holders of different forms of landed capital. Thus, a new social and cultural landscape based on the restructuring of landed capital has come into being. The core of the restructuring process is the creation of exchange value for amenity in rural property. This process creates status for property in land that is non-productive. It is part of a larger process of the reorganization of social reproduction, but it brings together the land of the countryside into a single market. The changes in the rural property system converge around new definitions of use and commodity value for land.

The central theoretical observation of this chapter is that the restructuring of landed capital has been a response to the ruralization process. Under this process, the traditional agricultural value for land has been isolated from industry and, because of its connection to nature, has been rendered economically vulnerable. The crisis of agriculture resulted in the relative depopulation of the countryside and the opening of that countryside to urban colonization. The inverse of ruralization has been the urbanization of the countryside, a process spurred by the development of residential amenity as commodity, and drawing urban capital into the countryside along with urban people.

What has not been explored are the dimensions of the reorganization of class interests in association with the reorganization of landed capital. The various permutations of class alliances have not been developed. And what about women and ethnic groups in the new countryside? The elements of the social geography have only been hinted at in this chapter. Finally, the contradictions of transformation from productive to reproductive uses in the transformation of landed capital need to be identified. What can be said, and it seems to explain a great deal of the exurbanization process, is that capitalism as a system is constantly restructuring its landed base. The transformation of the city's countryside is a major aspect of the creation of spaces under capitalism (Harvey, 1982).

References

Agricultural Council of Ontario (1986) *The Incomes of Farmers and Their Families in Ontario*. Agricultural Council of Ontario, Toronto.

Ambrose, P. (1974) *The Quiet Revolution: Social Change in a Sussex Village, 1871–1971*. Chatto and Windus, London.

Beesley, K.B. and Walker, G.E. (1990) Residence paths and community perception: a case study from the Toronto urban field. *The Canadian Geographer* 34, 318–30.

Bryant, C.R. and Russwurm, L.H. (1984) Changing population distributions and rural-urban relationships in Canadian urban fields. In: Bunce, M.F. and Troughton, M.J. (eds) *The Pressures of Change in Rural Canada*. Geographical Monograph 14, York University, Downsview, Ontario, pp. 113–37.

Bunce, M.F. (1981) Rural sentiment and the ambiguity of the urban fringe. In: Beesley, K.B. and Russwurm, L.H. (eds) *The Rural-Urban Fringe: Canadian Perspectives*. Geographical Monograph 10, York University, Downsview, Ontario, pp. 109–20.

Clout, H. (1972) *Rural Geography*. Pergamon, London.

Coppack, P. (1988) Reflections on the role of amenity in the urban field. *Geografiska Annaler* 70 B, 353–61.

Davies, S. and Yeates, M. (1990) Exurbanization as a component of migration: a case study of Oxford County, Ontario. *The Canadian Geographer* 35, 177–86.

Dorst, J.D. (1989) *The Written Suburb*. University of Pennsylvania Press, Philadelphia.

Durham, Regional Municipality (1991) *Subdivision Activity in the Rural Areas of Durham Region, Jan. 1, 1974 – December 31, 1989 Inclusive*. Planning Department, Whitby, Ontario.

Greater Toronto Coordinating Committee (1990) *Greater Toronto Area Urban Structure Concepts Study: Background Report No. 1 Description of Urban Structure Concepts*. IBI Group, Toronto.

Harvey, D. (1982) *The Limits to Capital*. University of Chicago Press, Chicago.

Joseph, A. and Smit, B. (1985) Rural residential development and municipal service provision: a Canadian case study. *Journal of Rural Studies* 1, 321–38.

Kantor, R. (1990) *Space for All: Options for a Greater Toronto Greenlands Strategy*. Queen's Printer, Toronto.

Lepine, Y. and Brunet, T. (1984) Les variations spatiales de la presence exurbaine en milieu rural quebecois: description et hypotheses d'explication. In: Bunce, M.F. and Troughton, M.J. (eds) *The Pressures of Change in Rural Canada*. Geographical Monograph 14, York University, Downsview, Ontario, pp. 138–59.

Lowe, P.D. (1977) Amenity and equity: a review of local environmental pressure groups in Britain. *Environment and Planning A* 9, 35–58.

Maitland, D. (1988) Relocation of farmers in the urban fringe. Unpublished BA thesis, York University, Toronto.

Marsden, T. (1989) Restructuring rurality, from order to disorder in agrarian political economy. *Sociologia Ruralis* 29, 312–17.

Marsden, T. and Murdoch, J. (1990) Restructuring rurality: key areas for development in assessing rural change. *Countryside Change Working Paper Series* 4, University of Newcastle upon Tyne, Newcastle upon Tyne.

McCrae, J.D. (1980) *The Influence of Exurbanite Settlement on Rural Areas: A Review of the Canadian Literature*. Working Paper 3, Lands Directorate, Environment Canada, Ottawa.

Michie, G.H. and Found, W.C. (1976) Rural estates in the Toronto region. *Ontario Geography* 10, 15–26.

Mitchell, R. (1989) *Canada's Population from Ocean to Ocean*. Ministry of Supply and Services Canada, Ottawa.

Momsen, M. F. (1984) Urbanization of the countryside in Alberta. In: Bunce, M.F. and Troughton, M.J. (eds) *The Pressures of Change in Rural Canada*. Geographical Monograph 14, York University, Downsview, Ontario, pp. 160–80.

Punter, J.V. (1974) Urbanites in the countryside: case studies in the impact of exurban development. Unpublished PhD thesis, University of Toronto.

Russwurm, L.H. (1976) Country residential development and the regional city form in Canada. *Ontario Geography* 10, 79–96.

Spectorsky, A. (1955) *The Exurbanites*. Lippincott, Philadelphia.

Statistics Canada (1974) *Perspective Canada*. Ministry of Supply and Services Canada, Ottawa.

Statistics Canada (1977) *Perspective Canada II*. Ministry of Supply and Services Canada, Ottawa.

Statistics Canada, Adler, H.T. and Brusegare, D.A. (eds) (1980) *Perspective Canada III*. Ministry of Supply and Services Canada, Ottawa.

Statistics Canada (1987) *Agriculture: Ontario*. Ministry of Supply and Services Canada, Ottawa.

Statistics Canada (1988a) *Population and Dwelling Characteristics, Census Metropolitan Areas and Census Agglomerations, Part 1*. Ministry of Supply and Services Canada, Ottawa.

Statistics Canada (1988b) *Census Tracts, Toronto, Part 1, Profiles*. Ministry of Supply and Services Canada, Ottawa.

Statistics Canada (1988c) *Census Tracts, Toronto, Part 2, Profiles.* Ministry of Supply and Services Canada, Ottawa.

Troughton, M.J. (1976) *Landholding in a Rural-Urban Fringe Environment: The Case of London, Ontario.* Occasional Paper 11, Lands Directorate, Environment Canada, Ottawa.

Walker, G. (1987) *An Invaded Countryside.* Geographical Monograph 17, York University, Downsview, Ontario.

Walker, G. (1988) *The Farmers' Class.* Geography Department Discussion Paper 34, York University, Downsview, Ontario.

Yeates, M. (1975) *The Windsor-Quebec City Urban Axis.* Macmillan, Toronto.

York, Regional Municipality (1991) *Subdivision Dwelling Unit Analysis.* Planning Department, Newmarket, Ontario.

4

A MEDIUM-TERM PERSPECTIVE ON DEMOGRAPHIC CHANGE IN THE AMERICAN MIDLANDS, 1803–1990

Clark Archer

Fast-driven pickup trucks with camper tops have replaced slower prairie schooners along the dusty roads, and huge self-propelled combines have taken the place of more human-sized horse-drawn reapers and stationary steam-driven threshers in the fields at harvest time. The dry, turbulent weather and shifting crop prices are staples of conversation, much as they were when pioneers first cut deep, long furrows into fertile prairie soil using John Deere's revolutionary steel moldboard plough. Each decennial census in the 19th century counted more people in the American Midlands. Even now, the latest census statistics are likely to be reported prominently by Midlands newspapers. Trumpeting early returns from the 1990 Census of Population, the Omaha, Nebraska *World Herald* (Ivey, 1991, p. B1) cheerfully blared that '100 Years After Heyday, Cass County Grows Again'. More sombrely, the Lincoln, Nebraska, *Journal Star* (Associated Press, 1990, p. 19) reported that 'US Farm Population Drops Again'.

The Midlands have been described poetically as 'vast and open, as if scaled for long-striding giants rather than for humankind' (Jones, 1968, p. 9). They occupy more than one-quarter of the land area of the contiguous United States. East to west they extend from the Mississippi River to the Rocky Mountains, and north to south from the US-Canadian border to the Kansas-Oklahoma border. The ten states of the Midlands study region together grow about half the corn, soyabeans and wheat harvested in the United States (US Bureau of the Census, 1989). They also produce more than one-third of US livestock products, as well as significant amounts of barley, oats, potatoes, rye, sorghum and sugarbeets. The dominance of agricultural land uses prompts a working hypothesis that alterations in agricultural technology or market conditions are apt to force changes in population distribution and settlement pattern.

The purpose of this study is to survey demographic changes in the American Midlands using a medium-term perspective. Temporally, this perspective encourages study of recent demographic changes in the light of earlier patterns and processes that exerted lasting impacts on the region's settlement structure. Spatially, this perspective encourages study of the Midlands as an open system, with demographic, economic and political linkages to the remaining parts of the United States and the rest of the world. Particular attention is focused on the region's rural farm population, but urban exchanges, migration flows, farm marketings, government programmes, and other public and private transactions that bind the Midlands into larger regional, national and global settlement systems continue to impact the demography of the region. The main objective is to achieve a better understanding of changing demographic patterns within the Midlands, including an assessment of the descriptive relevance of the recently asserted 'rural renaissance hypothesis'.

Nineteenth Century Demographic Changes

Profound demographic changes started when EuroAmericans began to settle the Midlands in the 19th century. When purchased from France in 1803 (Agnew, 1987, pp. 37–38), EuroAmerican settlement was limited to a narrow strip along the west bank of the Mississippi, between New Madrid and St. Louis (Semple, 1903, pp. 99 and 108). The prairies and plains of Upper Louisiana were occupied by semi-nomadic Indians, whose life-styles had been changed by the indirect introduction of the horse. The horse made it possible to hunt large game animals including American bison efficiently, and radically but briefly raised aboriginal living standards (Jones, 1968, pp. 31–42; Brown, 1985, p. 50). The famed Lewis and Clark expedition of 1804 also found subsistence agricultural villages along the Missouri River and its tributaries, occupied by such tribes as the Otoe, Omaha and Ponca (Garrett, 1988, pp. 34–35).

After 1800, the Native American population diminished sharply through disease, warfare and forced relocation. For example, many of the Pawnee Nation of Nebraska perished in the cholera epidemic of 1849, infected by 'Forty-niners' rushing along the Oregon Trail to the California goldfields. The tribe diminished from over 12 000 in 1838 to under 3500 by 1861 (Federal Writer's Project, 1939, p. 29). In very recent years, however, the Native American population has again increased. In 1990, the US Census counted about 275 000 American Indians in the Midlands, comprising 1.2% of the study area's population – twice the proportion for the US as a whole (US Bureau of the Census, 1991b, p. 3).

In 1860, the Midlands' EuroAmerican population stood at 2.2 million, or about 7% of that of the United States as a whole (Table 4.1). Then a tide of 2.5 million domestic and foreign immigrants poured into the region during the last decades of the 19th century. One-fifth of the Midlands'

Table 4.1. Midlands population by state, 1860–1920.

Area	Admission	Population (Thousands of persons)						
		1860	1870	1880	1890	1900	1910	1920
Colorado	1876	34	40	194	413	540	799	940
Iowa	1846	675	1194	1625	1912	2232	2225	2404
Kansas	1861	107	364	996	1428	1470	1691	1769
Minnesota	1858	172	440	781	1310	1751	2076	2387
Missouri	1821	1182	1721	2168	2679	3107	3293	3404
Montana	1889	–	21	39	143	243	376	549
Nebraska	1867	29	123	452	1063	1066	1192	1296
North Dakota	1889	–	2	37	191	319	577	647
South Dakota	1889	5	12	98	349	402	584	637
Wyoming	1890	–	9	21	63	92	146	194
Midlands		2204	3926	6412	9551	11 223	12 959	14 227
Density/sq.mi.		2.5	4.5	7.4	11.0	13.0	15.0	16.4
% of US popn.		7.0	10.2	12.8	15.2	14.7	14.1	13.4
US		31433	38 558	50 189	62 980	76 212	92 228	106 022
Density/sq.mi.		10.6	10.9	14.2	17.8	21.5	26.0	29.9

Table 4.1. (contd.) Midlands population by state, 1930–1990.

Area	Admission	Population (Thousands of persons)						
		1930	1940	1950	1960	1970	1980	1990
Colorado		1034	1123	1325	1754	2210	2890	3294
Iowa		2471	2538	2621	2758	2825	2914	2777
Kansas		1881	1801	1905	2179	2249	2364	2478
Minnesota		2564	2792	2982	3414	3806	4076	4375
Missouri		3629	3785	3955	4320	4678	4917	5117
Montana		538	559	591	675	694	787	799
Nebraska		1378	1316	1325	1411	1485	1570	1578
North Dakota		681	642	620	632	618	653	639
South Dakota		693	643	653	681	666	691	696
Wyoming		226	251	291	330	332	470	454
Midlands		15 095	15 250	16 268	18 153	19 563	21 330	22 207
Density/sq.mi.		17.4	17.6	18.8	20.9	22.6	24.6	25.6
% of US popn.		12.3	11.5	10.8	10.1	9.6	9.4	8.9
US		123 203	132 165	151 326	179 323	203 302	226 543	248 710
Density/sq.mi.		34.7	37.2	42.6	50.6	57.5	64.0	70.3

Sources: US Bureau of the Census, *Census of Population*, US Government Printing Office, Washington DC, various dates; US Bureau of the Census, *Census and You*, Vol. 26, No. 4, (1991b, pp. 3–4).

population was foreign born in 1890, and the proportion was substantially higher along the advancing settlement frontier (Semple, 1903, pp. 312–17; Garrett, 1988, pp. 52–3). By 1890, the region contained 9.6 million, or over 15% of the US population.

Unfortunately, many new settlers faced unexpectedly severe conditions. What are now known to have been unusually wet years were succeeded by drought in the 1890s. Crops withered and livestock prices plummeted as frightened ranchers sold or slaughtered herds they could no longer feed. Lack of rain forced a 'deluge of mortgage foreclosures' (Porter, 1989, p. 3). Thousands of prairie schooners stirred up dust in an ebb tide of out-migration, and vast expanses of the Plains from Kansas to North Dakota became almost depopulated near the turn of the century.

A number of counties in the Midlands reached their largest Euro-American population by 1890 or before. Most of these counties are located in eastern Iowa, central Missouri, eastern Kansas, or eastern Nebraska (Fig. 4.1). Usually within easy access to water transport along the Missouri or Mississippi Rivers, these counties were often the first to be opened to homesteading. Their initial booms soon fizzled out, however, and they have never again been so populous as they were a century or more ago.

Cass County, Nebraska, positioned at the confluence of the Platte and Missouri Rivers, is an example. The county's population grew dramatically from 3369 in 1860 to a peak of 24 080 in 1890, before sliding to a low of 16 361 in 1940. More recent growth to 21 318 in 1990 prompted the newspaper headline cited above, and can be attributed to exurban spillover from nearby Omaha and Offutt Air Force Base, headquarters of the Strategic Air Command. Most Midlands counties, which peaked in 1890 or before and which are not now near metropolitan centres, have failed to evidence recent population growth.

Early Twentieth Century Demographic Changes

The Twentieth Century managed to open prosperously in the Midlands. New strains of wheat introduced from Russia, improved dry farming methods, years of normal to above normal precipitation, and soaring food prices during the First World War boosted farm earnings to unprecedented levels. Farmers' prosperity in turn contributed to the revenues of merchants, brokers and lenders in market towns, and to the wages and profits of workers and manufacturers busy building farm equipment in local industrial centres such as Iowa City, Iowa, where the first petrol-powered tractor was built. There was as much as a one-quarter increase in the size of the Midlands population, to over 14 million by the end of the second decade of the 20th century. Yet population growth in the region had already begun to lag behind that of the nation for, in comparison, the

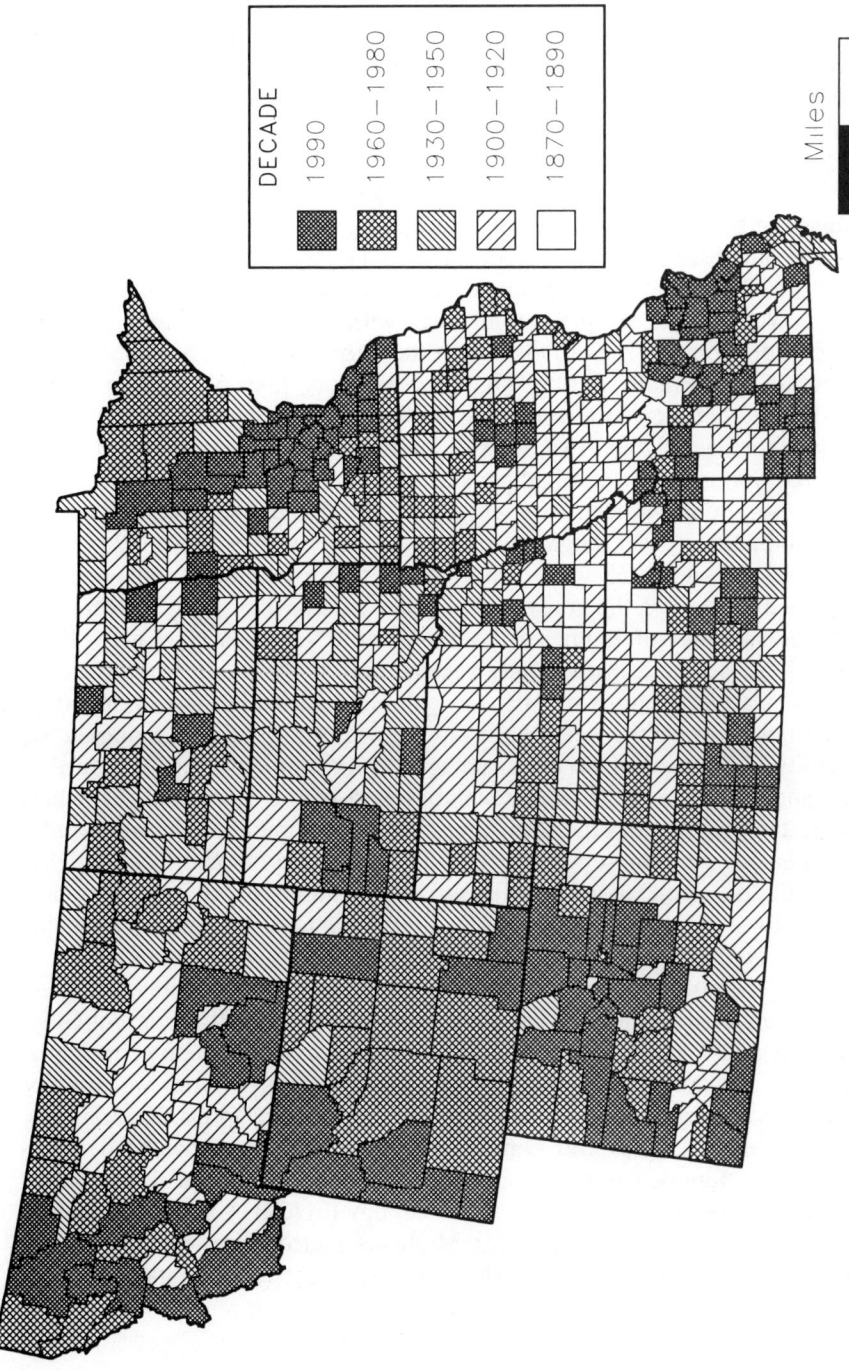

Fig. 4.1. Decade of maximum population.

population of the US as a whole expanded by two-fifths from 1900 to 1920. And the Great Depression was just around the corner, a period when all ten Midlands states suffered out-migration.

Agrarian distress was a prime cause of population increase being nearly reversed by out-migration. The price of wheat per bushel plummeted from US$1.44 in 1925 to US$.67 in 1930, before dropping further to less than US$.40 in 1931 (US Bureau of the Census, 1960, pp. 296–97). The market values of other crops also declined, dropping the index of farmland value, which had reached 173 in 1920, back to 114 by 1930 (Lee and Passell, 1979, p. 365). These calamitous conditions pushed many farmers off the land. Yet despite movement to cities within the region, and net out-migration from the region, the size of the farm population of the Midlands increased marginally during the 1920s, to more than 6.2 million by the end of the decade. As a result, the Midlands' farm population reached its maximum size in 1930, two decades later than the farm population of the United States as a whole had peaked.

Farm population density averaged 3.9 persons per square kilometre in the Midlands in 1930. Higher farm population densities above 6.2 persons per square kilometre were achieved by many counties in the eastern cornbelt section of the study area, which extends over most of Iowa and nearby subareas of Minnesota, Missouri and Nebraska (McCarty, 1940, p. 295). Elsewhere in the study region, only the alluvial 'Bootheel' and Ozark plateau sections of Missouri exhibited such high farm population densities. Toward the west, there was a downturn in rural farm population densities, in response to diminishing average rainfall levels and further distances from markets. Many counties west of the 51 cm isohyet, which almost equally divides the study area, had farm population densities under 1.5 persons per square kilometre in 1930. However, some locally higher densities occurred where the Missouri or its tributaries provided an opportunity for irrigated cultivation, as an alternative to dry farming or ranching (McCarty, 1940, p. 231).

Fewer than half the 761 counties in the ten Midlands states contained an urban settlement of 2500 persons or more in 1930. Consequently, although 42% of the region's total population lived in urban centres, the average at county-level was merely 17%. Most large cities were major river ports or rail centres, including St. Louis, Minneapolis, Kansas City, Denver, Des Moines and Omaha. But many smaller urban settlements had been 'founded not in response to economic or social needs but in anticipation of such needs' (Jones, 1968, pp. 131–32). Unmet anticipation caused scores of once hoped for Midlands metropolises to fail to grow larger than hamlets, especially those unable to get an early start as a port along one of the region's larger rivers.

Mid-twentieth Century Demographic Changes

Many important changes in the characteristics and geographical distribution of population during the next half-century sprang from changes in farm technology and agricultural market structure. Regarding market structure, one-fifth of US harvested cropland was devoted to growing fodder for 20 million draft animals as late as 1930 (US Bureau of the Census, 1960, p. 281, 289). By 1987, most of the 2 million horses and ponies – mules were no longer counted – were used for recreation (US Bureau of the Census, 1989, p. 25). Conversely, the proportion of US harvested cropland devoted to export production expanded from one-tenth in 1930 to over one-third in 1980. Export production became especially important to cash-grain farmers, since one-third of the US corn and soyabean crops and nearly two-thirds of the US wheat crop were sold to foreign buyers in 1980.

Farm technology also changed as American agriculture became highly mechanized (Rasmussen, 1982; Hart, 1986, 1987). In 1930, when the rural farm population of the Midlands reached its peak, barely one-quarter of Midlands farms used tractors. At that time, it required an average of nearly 190 hours of farm labour to produce 100 bushels of corn and 100 bushels of wheat. By 1950, nearly three-quarters of Midlands farms used tractors, and nine-tenths did so by 1980. Other changes included higher-yielding hybrid plants, and increased applications of fertilizer and other chemicals. As a result, 100 bushels of corn and 100 bushels of wheat could be produced using less than 15 hours of labour in 1980. By implication, labour productivity rose about 1330% on Midlands cash-grain farms between 1930 and 1980.

Agricultural mechanization impacted the number of farm units much more than the amount of land used for agricultural purposes in the Midlands. As a result, while total farm area expanded marginally from 147 to 166 million ha, the total number of farm units plummeted from 1.2 million in 1930 to 622 000 in 1980. These changes meant that the average size of a Midlands' farm rose from 119 ha to 267 ha between 1930 and 1980.

Nearly 6.3 million people had lived on Midlands farms and ranches in 1930. But increases in farm worker productivity due to mechanization so diminished the agricultural work-force that by 1980 the region's farm population shrank to less than 1.7 million, or only 7.9% of the Midlands' total population. Notably, the decline in the size of the farm population was proportionally greater than the decline in the number of farm units, because the average number of occupants per farm dropped from 5.06 in 1930 to 2.72 in 1980.

Larger, more powerful and more numerous farm machines led to fewer farm workers and smaller farm families. Overall, there were 73% fewer residents on Midlands farms and ranches in 1980 than in 1930, despite little change in the total area of farmland. Between 1930 and 1980, the region-wide average rural farm population density dropped from 3.9 to just 1.3 persons per square kilometre, corresponding to about one farmstead per section of land by the later year. The relative pattern of farm settlement remained similar, with higher densities in the east and lower densities in the west, but no Midlands county had a rural farm population density as high as 4.6 persons per square kilometre in 1980. In the western rangelands, densities fell below 0.4 farm residents per square kilometre, lending statistical support to a University of Nebraska student's lament that 'Now I have to travel over a mile to the nearest neighbors' (Dauel and Beals, 1988, p. 3).

In addition, many small hamlets and towns also suffered decline, for as the populations of their rural hinterlands diminished there was less local demand for the goods and services that these settlements traditionally had provided. Nevertheless, although the region still remained less urbanized than the United States as a whole, nearly two-thirds of all Midlanders lived in cities or metropolitan areas by 1980.

Rural Renaissance Reversal

After the 1980 census results appeared, several observers attested that a 'rural renaissance' (Roseman, 1985, pp. 49–52) or 'nonmetropolitan turnaround' (Wardwell and Brown, 1980, pp. 16–20) had occurred. Wardwell and Brown (1980, p. 12) went so far as to assert that 'Past notions of metropolitan concentration and of decline or abandonment of small towns and rural areas can no longer serve as guides for understanding population distribution in the United States'. In the Midlands, North and South Dakota both increased in population for the first time since the Depression. However, only three Midlands states – Colorado, Montana and Wyoming – had more in-migrants arriving than out-migrants leaving during the 1970s. Yet the strength of the turnaround was muted in the Midlands. Nationally, nearly three out of four non-metropolitan counties grew in population during the 1970s, but in the Midlands barely half the non-metropolitan counties managed to achieve population gains. Nonetheless, there was a modest turnaround, since three out of four non-metropolitan counties in the region had suffered population losses as great as 70% between 1930 and 1970. The largest county-level loss was less than 25% from 1960 to 1970, and Midlands non-metropolitan counties grew by an average of nearly 7% during the decade.

Unfortunately, the results of the 1990 census enumeration indicate that the rural renaissance went bust after 1980 in the Midlands. On average, Midlands non-metropolitan counties suffered a 5.1% decline in population from 1980 to 1990. In comparison, the region's metropolitan counties gained an average of 12.3%. Although nearly four out of five metropolitan counties in the region gained population between 1980 and 1990, the corresponding proportion was less than one out of four among non-metropolitan counties.

In order to examine the trends more closely, Midlands counties were grouped according to their population losses or gains during three time periods: 1930–1970, 1970–1980, and 1980–1990 (Table 4.2, Fig. 4.2). These times bracket the assumed rural renaissance era for comparison with earlier and later periods. Evidence of a 1970s rural renaissance is shown by the fact that 31% of Midlands non-metropolitan counties advanced from being population losers between 1930 to 1970 to being population gainers between 1970 and 1980. Another 25% of non-metropolitan counties gained population during both of these intervals. But whereas only 4% of Midlands non-metropolitan counties slid from gainers to losers between 1930 and 1970 and 1970 and 1980, 40% were losers during both periods, indicating that the effects of the rural renaissance were not enjoyed uniformly throughout the region.

In the later periods, 42% of Midlands non-metropolitan counties lost population from 1970–1980, and from 1980–1990. Many of these counties actually continued population loss trends that had begun earlier. Indeed,

Table 4.2. Frequencies of counties by population change classes for Midlands non-metropolitan and metropolitan counties, 1930–1990.

Time period		Non-MSA		MSA[a]		All	
1930–1970	1970–1980	No.	%	No.	%	No.	%
Loss	Loss	273	40	2	3	275	36
Gain	Loss	30	4	8	10	38	5
Loss	Gain	208	31	4	5	212	28
Gain	Gain	172	25	64	82	236	31
	Total	683	100	78	100	761	100
1970–1980	1980–1990						
Loss	Loss	288	42	6	8	294	39
Gain	Loss	231	34	10	13	241	32
Loss	Gain	15	2	4	5	19	2
Gain	Gain	149	22	58	74	207	27
	Total	683	100	78	100	761	100

[a]Metropolitan Statistical Area.
Data Sources: US Bureau of the Census, *Census of Population*, US Government Printing Office, Washington DC, various dates; US Bureau of the Census, *County and City Data Book, 1988 on diskettes [Machine-readable data]*, US Bureau of the Census, Washington DC; US Bureau of the Census, *Census of Population and Housing, 1990: Public Law (P.L.) 94–171 Data on CD-ROM [Machine-readable data]*, US Bureau of the Census, Washington DC.

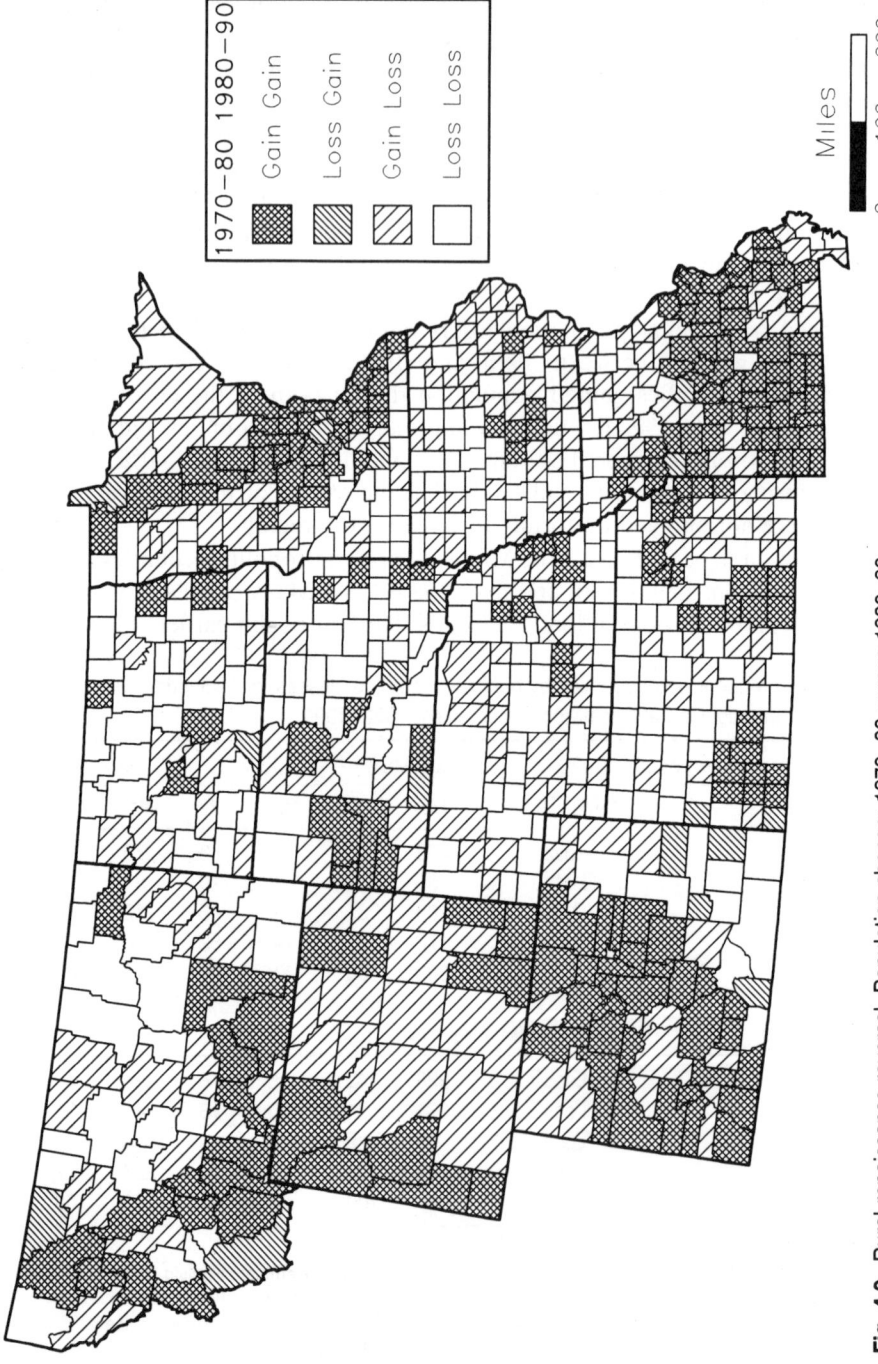

Fig. 4.2. Rural renaissance reversal. Population change: 1970–80 versus 1980–90.

over one-third of all counties in the study area reached their maximum populations before 1930, and nearly two-thirds did so before 1960. While 22% of Midlands non-metropolitan counties increased in population from 1970–1980 and from 1980–1990, only 2% rose from losers to gainers between these two time intervals. Yet the reversal of the rural renaissance is most apparent in the 34% of Midlands non-metropolitan counties which slid from being gainers during the 1970s to being losers during the 1980s.

Strong pressures toward urbanization are evidenced by the fact that three-quarters of the region's metropolitan counties grew in population over each of the three time periods, 1930–1970, 1970–1980, and 1980–1990. Hence, metropolitan counties generally outperformed non-metropolitan counties in the Midlands even during the rural renaissance decade of the 1970s. Moreover, most of those rare non-metropolitan counties that managed to advance from losers in the 1970s to gainers in the 1980s are to be found within commuting distance of metropolitan employment opportunities. Consequently, counties grouped in the 'gain–gain' and 'loss–gain' categories typically are located within or adjacent to larger metropolitan areas, including St. Louis, Kansas City, Minneapolis-St. Paul and Denver. Others contain the central cities or suburbs of smaller metropolitan areas, such as Wichita, Omaha, Des Moines or Lincoln.

Most remaining 'gain–gain' counties are found in the Ozark uplands of Missouri, or the Rocky Mountain highlands of Montana, Wyoming, or Colorado. It should be noted, however, that the Rocky Mountains highlands along the western edge of the study area are still lightly settled, often with total population densities of 1.5 persons per square kilometre or less in 1990 (Fig. 4.3). Small numbers of in-migrants can prompt seemingly large gains in the Rockies. Although traditionally lagging, the scenic Missouri Ozarks recently have become appealing as a comfortable setting for affordable retirement living (Reeder and Glasgow, 1990).

The largest county groupings, which comprise the 'loss–loss' and 'gain–loss' categories for the 1970–1980, and 1980–1990 time periods, hold two-thirds of all counties in the study area. Counties that suffered losses during both recent decades were especially common within a trapezoidally shaped zone which extends from central Iowa westward across the Dakotas, Nebraska and Kansas. The edges of this zone of population loss during each of the last two decades appear to coincide with the outer boundaries of the agriculturally defined cornbelt and the spring wheat region, where large fractions of the US corn, wheat and soyabeans crops are grown. Statistical tests using correlation and regression analysis of the strength of the apparent geographical correspondence between productive agricultural land use and troubling population decline trends are reported in this chapter.

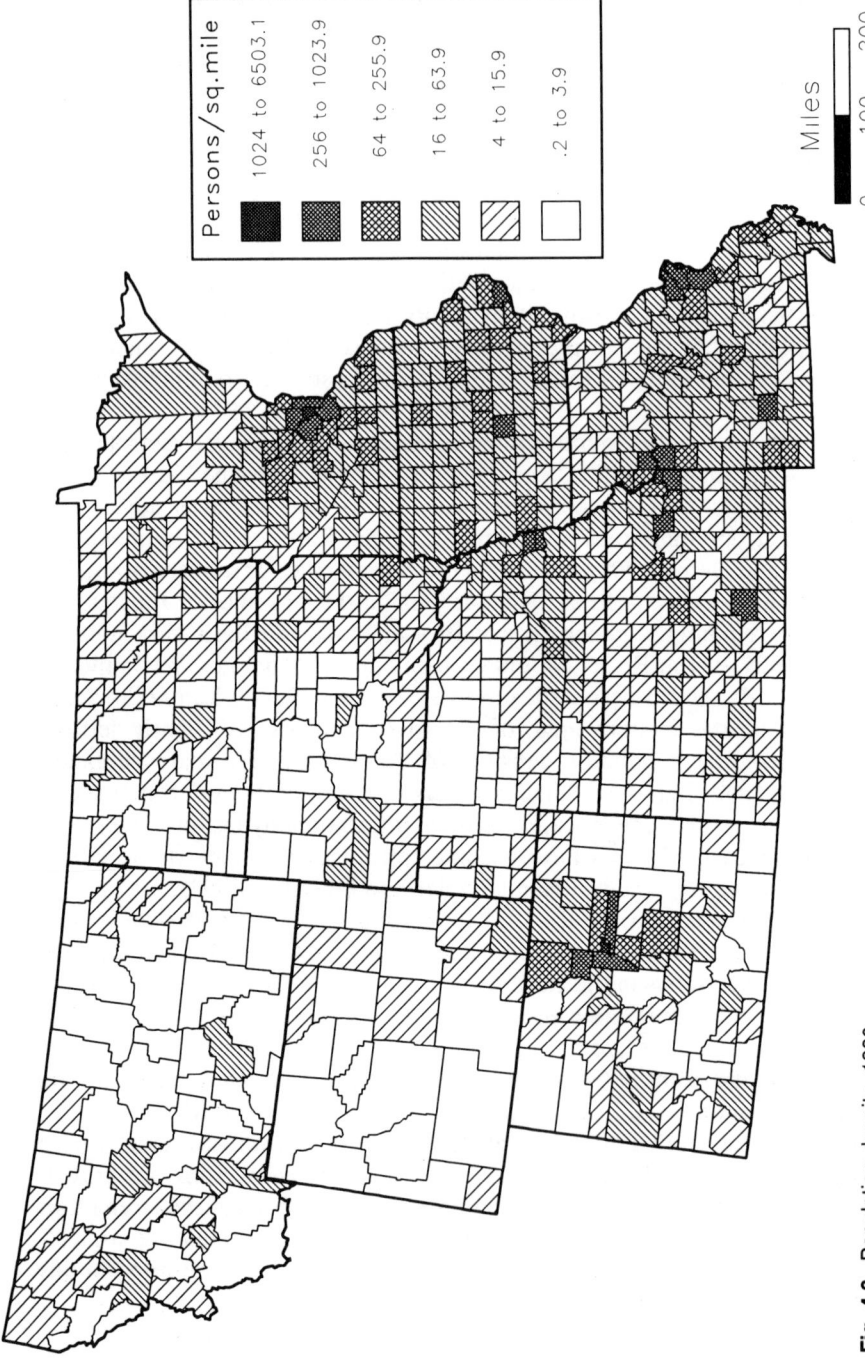

Fig. 4.3. Population density, 1990.

Agricultural Restructuring During the 1980s

The major causes of agrarian distress during the 1980s once again were beyond the reach of Midlands farmers and ranchers. Drought exacted a toll as another cycle of low rainfall afflicted the region. Serious droughts in 1983 and 1986 were preludes to the drought of 1988, which was one of the most severe on record (Dyson, 1989, p. 7). The effects were most intense in the eastern part of the region, where fewer farmers have invested in irrigation equipment because of normally higher rainfall. In the west, most fields used for crops, including hay, are irrigated even in moist years.

But dry weather was not the worst of Midlands farmers' woes. Softening foreign demand for US farm products was even more distressing. The US exported 35% of corn production, 40% of soyabean production, and 64% of wheat production in 1980. Midlands farmers were euphoric; so were their creditors. Then came the crash. Five years later, the US exported just 14% of corn production, 36% of soyabean production, and 38% of wheat production. Federal agricultural assistance payments reached over US $30 billion, but Midlands farm income still plummeted.

Land values are a sensitive indicator of agricultural well-being. In 1982, farmland and ranchland were worth an average of almost US $1730 per ha in Midlands non-metropolitan counties. By 1987, agricultural land values had slipped 28%, to an average of US $1260 per ha in Midlands non-metropolitan counties. The largest declines were suffered in cornbelt sections of Iowa, Minnesota and Nebraska, where many counties experienced agricultural land-value declines of 40% or more.

As the market worth of their landholdings shrunk below the burdens of their accumulated debts, thousands of previously 'cash poor but land rich' farmers suffered foreclosure. Many moved off to seek work in town, some retired to make-do on food stamps and Social Security cheques, and a few became part-time tenant farmers cultivating lands that they had once owned. The well attended 'FarmAid' concerts promoted by singer Willy Nelson generated some solace, but little else. The roll-out effects of the farm crisis into nearby towns are crudely but effectively conveyed by noting that one-third of farm equipment dealers in Iowa and Nebraska went out of business in the mid–1980s (Jordon, 1988, p. 1-M).

Nevertheless, it is important to observe that the farm recession of the 1980s caused very little change in the geographical extent of agricultural land use, at least within the study area. In the Midlands, the total area of land devoted to agriculture declined by less than 5%, from 166 million ha in 1980, to 159 million ha in 1987. However, much farmland changed ownership, as a diminishing number of more fortunate farmers took advantage of unbelievable bargains to increase their scales of operation. As a result, Midlands farms expanded to an average size of 284 ha by

1987. More intensive mechanization made such expansion possible. In 1987, Midlands farmers needed an average of just 10 hours of work to plant, grow and harvest 100 bushels of wheat and 100 bushels of corn. In 1930, as previously noted, nearly 190 hours of labour effort had been needed to achieve the same output. Agricultural mechanization, which includes machinery, chemicals, and hybrid plant genetics, is among the root causes of recent agrarian restructuring in the American Midlands.

Correlates of Population Change During the 1980s

Correlation and regression analyses were undertaken using county-level data in order to examine the underlying covariates of population change in the Midlands during the 1980s. The chosen research variable is county-level percentage population change from 1980 to 1990 (Fig. 4.4). Since rural population change is the main focus, counties were divided between non-metropolitan and metropolitan categories using 1980 Metropolitan Statistical Area (MSA) demarcations. Potentially related explanatory variables were selected on the basis of background research and availability of data in machine-readable form from US Census sources (US Bureau of the Census, 1988, 1990b, 1991a). Many of the chosen explanatory variables pertain to county-level agricultural conditions, such as farm population density or percentage of farms receiving Commodity Credit Corporation (CCC) loans. Others are indicators of general economic and demographic conditions, such as income per capita or percentage urbanization.

Bivariate Pearson product-moment correlation coefficients were computed using 1980–1990 percentage population change as the dependent variable, and the selected independent variables for non-metropolitan, metropolitan, and all counties in the Midlands study area. Only one of the correlation coefficients was found to exceed 0.50, and many are less than 0.25 in value (Table 4.3).

The modest sizes of most computed correlation coefficients suggests that a simple, one variable based explanation is unlikely to be found for county-level variations in population change within the study area. Nevertheless, it is noteworthy that among non-metropolitan counties variations in percentage population change tend to correlate more strongly with agricultural indicators than with the selected general economic indicators. For example, correlations are found to be higher for percentage farm population and percentage area in farms, than for income per capita or percentage unemployment. Other comparatively strong correlations with variations in percentage population change among non-metropolitan counties are encountered for percentage of: full-time farmers; farms receiving government payments, and farms receiving CCC loans.

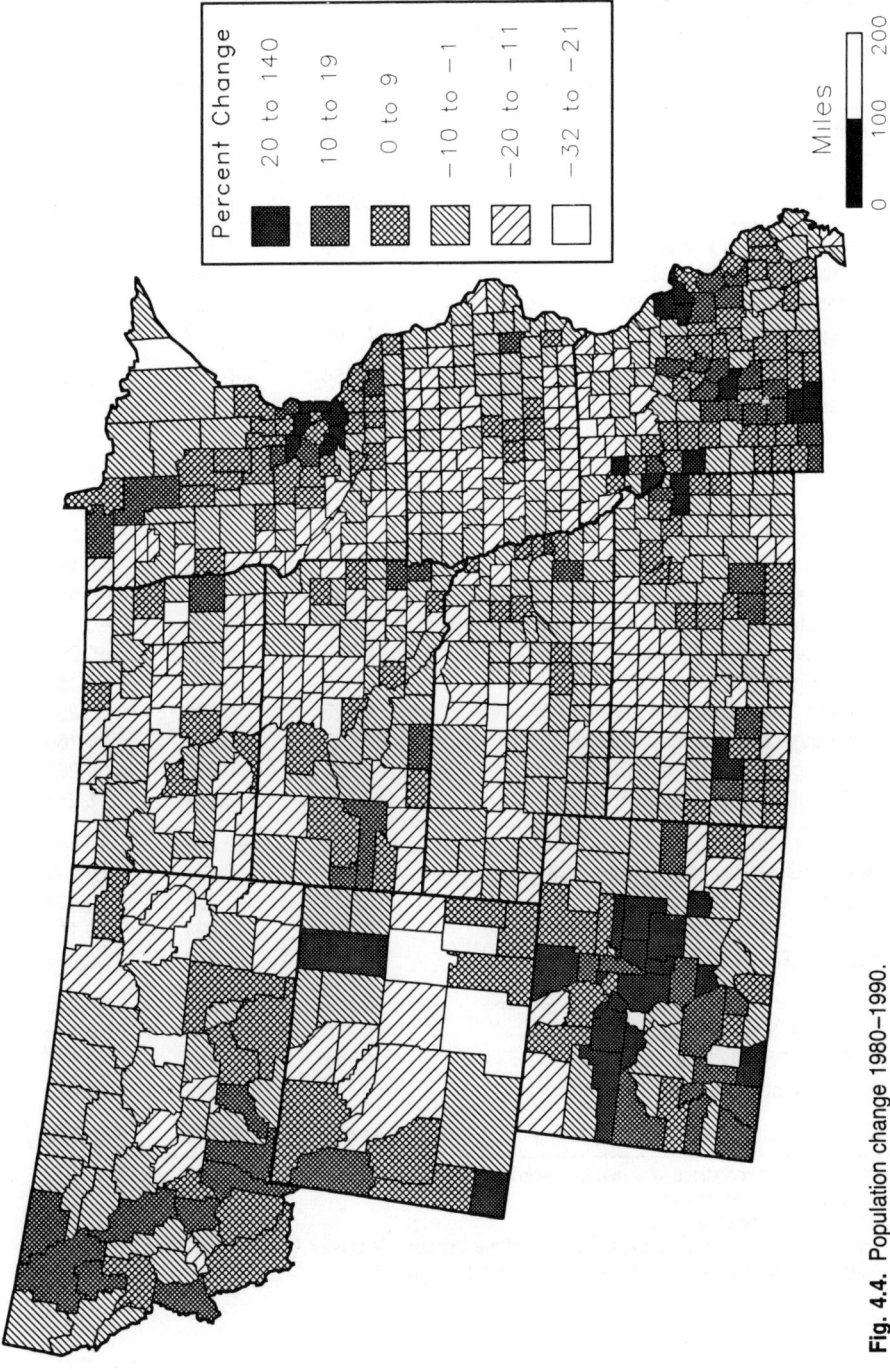

Fig. 4.4. Population change 1980–1990.

Table 4.3. Selected 1980–1990 percentage population change correlations, for Non-MSA, MSA and all Midlands counties.

Dependent Variable: Percentage population change, 1980–1990

Independent variable	Pearson Product Moment Correlation County Grouping		
	Non-MSA	MSA[b]	All
Population Density 1990	0.256	−0.174	0.082
% Urban Population 1980	0.126	−0.288	0.232
% Population Under 18 1990	0.113	0.320	0.124
% Population White 1990	−0.149	0.242	−0.115
Farm Population Density 1980	−0.054	−0.069	0.021
% Farm Population 1980	−0.475	−0.044	−0.497
% Full-time Farmers 1987	−0.451	−0.261	−0.497
Median Age of Farm Operators	0.255	0.141	0.238
Number of Farms 1987	0.021	−0.104	0.091
% Change Number of Farms 1978–87	0.248	0.420	0.240
Average ha per Farm 1987	−0.079	−0.168	−0.132
% Farms over 202 ha 1987	−0.263	−0.109	−0.320
% of Total Area in Farms 1987	−0.420	−0.206	−0.407
% of Total Area in Cropland 1987	−0.319	−0.247	−0.294
% Change in Cropland 1982–87	0.242	0.074	0.159
Average Value Farmland per ha 1987	0.125	0.106	0.257
% Change Avg. Val./ha 1978–82	0.127	0.512	0.169
% Change Avg. Val./ha 1982–87	0.281	0.253	0.283
% Farm Marketings from Crops 1987	−0.328	−0.082	−0.255
% Farms with Govt. Payments 1987	−0.472	−0.240	−0.460
Average Govt. Payment Per Farm 1987	−0.217	−0.257	−0.249
% Farms with CCC Loans 1987	−0.421	−0.235	−0.409
Average CCC Loans 1987	−0.086	−0.224	−0.113
% Land in Diversion Programmes 1987	−0.331	−0.302	−0.337
Income Per Capita 1984	−0.107	0.324	0.035
% Income from Manufacturing 1984	0.170	0.068	0.198
% Below Poverty Line 1979	−0.141	−0.363	−0.251
% Unemployment 1986	0.132	−0.379	0.023
Number of County-level Observations[a]	589–682	67–78	660–760

[a]Numbers of observations vary due to 'disclosure rule' restrictions on data indicative of specific individuals, families, or firms.
[b]Metropolitan Statistical Area
Data Sources: as Table 4.2 and US Bureau of the Census, *Census of Agriculture, 1987 on CD-ROM [Machine-readable data]*, US Bureau of the Census, Washington DC.

Among metropolitan counties, in marked contrast, percentage population change tends to be more strongly associated with several of the selected general social and economic indicators. Within the metropolitan county grouping, percentage population change 1980–1990 correlates rather strongly with income per capita, percentage unemployment, and percentage population under 18 years of age. Additional comparatively high correlations are found for percentage change in number of farms and percentage change in agricultural land value per ha, although these statistical relationships seem to reflect the effects of land use conversion processes along the suburban fringes of counties in the metropolitan category.

The composite impression created by all the correlation coefficients considered together is that agricultural activities and population losses tended to be geographically coextensive in the Midlands during the 1980s. This finding adds weight to the working hypothesis of linkages between population decline and agricultural restructuring. Some puzzling coefficients can be noted, however, including the negatively signed correlations involving indicators of participation in federal agricultural assistance programmes and percentage population change 1980–1990 among counties in the non-metropolitan group. These coefficients suggest that higher federal farm assistance outlays and greater population losses are geographically coincident. Sometimes, such confusing anomalies can be untangled by using multivariate rather than bivariate tools of analysis.

Application of step-wise multiple regression had the twofold objective of simultaneously paring the rather long list of potential explanatory variables, and of ascertaining which statistical associations remained robust even after controlling for the statistical influences of other variables. Parameters were estimated for several dozen multiple regression models, using SYSTAT's Multiple General Linear Hypothesis (MGLH) procedure (Wilkinson, 1990) on a PC-DOS microcomputer. The final model specifications were arrived at through computationally constrained induction, a trial-and-error process leading to best-fit equations for the data examined. Some variables were 'forced' into the equations as controls, but others, which had been expected to be influential, were eliminated on statistical grounds of negligible linear partial correlations with the dependent variable, which was county-level percentage population change from 1980 to 1990. The search process was directed at deriving the statistically best equation for the non-metropolitan group of counties. It is plausible that statistically superior models could be estimated later, after all 1990 census data finally become available at county-level. Nevertheless, it can be pointed out that in the case of the final multiple regression model for non-metropolitan counties which is described, the overall equation and all included variables meet conventional statistical significance criteria (Table

Table 4.4. Standardized coefficient multiple regression models for percentage population change 1980–1990, Midlands Non-MSA and MSA Counties.

Dependent Variable: Percentage Population Change 1980 to 1990

| Independent Variable | Standardized Multiple Regression Coefficients | | | |
| | Non-MSA model | | MSA model | |
	Beta	T-value	Beta	T-value
% Urban Population 1980	−0.148	3.557*	−0.844	5.231*
% Population White 1990	−0.193	5.555*	−0.024	0.235
Farm Population Density 1980	0.200	4.766*	−0.047	0.408
% Farm Population 1980	−0.365	6.908*	−0.517	2.987*
% Full-time Farmers 1987	−0.114	2.357*	0.187	1.486
% Change Number of Farms 1978–87	0.195	4.451*	0.401	3.483*
Average ha per Farm 1987	−0.156	3.522*	−0.212	2.060*
% of Total Area in Farms 1987	−0.250	5.347*	−0.045	0.332
% Change in Cropland 1982–87	0.114	3.549*	−0.082	0.862
% Farm Marketings from Crops 1987	−0.111	3.016*	−0.026	0.248
Income Per Capita 1984	0.087	2.314*	0.358	3.642*
% Unemployment 1986	−0.158	4.308*	−0.288	3.277*
Multiple R-squared	0.457		0.649	
F-ratio	45.569*		9.696*	
Number of Observations	664		76	

*Significant at 0.01 level of probability.
Data Sources: as Table 4.3.

4.4). However, not all the included variables are statistically significant in the case of the comparison model for metropolitan counties.

A multiple R-squared value of 0.457 indicates that the final model accounts for slightly less than half the total county-level variance in percentage population change from 1980 to 1990 among non-metropolitan counties in the Midlands. Although slightly more than half the total variance remains unaccounted for by the regression equation, the standardized partial regression coefficients (Betas) yield insights that were unattainable through bivariate correlation analysis alone. The largest standardized partial regression coefficients are for percentage farm population and percentage of total area in farms. Both coefficients are negative. The magnitudes of these coefficients indicate that even after statistically accounting for the effects of control variables entered in the final equation, including percentage white population, income per capita, and percentage unemployment, these farm settlement variables continue to exhibit the strongest statistical links to population change among non-metropolitan counties in the study region. There are also other indicators of the impacts of agricultural restructuring on non-metropolitan population change, including negative standardized partial regression coefficients for other variables involving percentage full-time farmers, average size of farm, and percentage farm marketings from crops. It is worth noting that the percentage farm marketings from crops variable, statistically edged federal farm assistance programme variables out of the final equation. Although scarcely confirmatory evidence, this does suggest a *post hoc* hypothesis that federal farm assistance efforts are correlated with rural population decline at county level, because the patterns of federal farm assistance and rural population decline in turn are both responsive to the consequences of agricultural restructuring.

Unexpectedly, the regression model estimated for metropolitan counties exhibits a higher R-squared, 0.649, than the regression model estimated for non-metropolitan counties. Apparently, variables chosen to unravel agrarian distress in rural and small town settings also highlight suburban expansion at the edges of metropolitan areas. Thus, for example, the oddly high negative standardized partial regression coefficient for percentage urban population becomes plausible after noting that proportionally high metropolitan population growth tends to occur where non-urbanized land still remains available for development. In a somewhat related fashion, the high coefficient for percentage change in number of farms is likely an artifact of the official census definition of a farm unit, as one with just US $1000 or more in agricultural sales. Accordingly, suburbanites who sell a couple of foals from their children's ponies, or a few bushels of apples from their hobby orchards, become statistical farmers.

The comparison model for metropolitan counties was not intended to be a best-fit model of metropolitan population growth; it was intended as a control comparison for the non-metropolitan model. In this capacity it performs effectively, showing that general social and economic indicators, such as percentage unemployment or income per capita, are closely related to county-level variations in population change for Midlands metropolitan counties. This contrasts sharply with non-metropolitan counties in the Midlands, among which variables measuring various aspects of agricultural conditions were found to be much more statistically reliable predictors of county-level variations in population change from 1980 to 1990, than were variables measuring general social or economic conditions, such as percentage unemployment or income per capita.

Conclusions

Looking back over two centuries of EuroAmerican presence in the American Midlands, it has become apparent that demographic change has not been unusual. In some parts of the region, demographic change was most profound in the 19th century, since a number of counties in the Midlands never again matched the population sizes and densities they achieved in the first boom of pioneer settlement. Many of these counties are situated along the Mississippi or Missouri Rivers at the eastern edge of the study area. In other parts of the region, demographic change was most severe during the dust bowl days of the 1930s, when a large number of counties in the central Midlands reached their maximum populations. In some Midlands counties, however, the most painful demographic changes did not occur until the latest farm crisis of the 1980s. Many of the counties whose populations appear to have crested during the last decade or two are located in the wetter, more fertile cornbelt subsection of the study region. Indeed, the statistical results of applying correlation and regression methods to county-level population changes between 1980 and 1990 show that rural population decline tended to be most severe in settings where agricultural potential was highest during the last decade. Unfortunately, Cass County-style rebounds have proven to be rare in the Midlands.

Agricultural restructuring has been driven by technological changes leading to more intensive agricultural mechanization, with the result that fewer and fewer farm owners and farm workers have been needed to produce the same or even greater quantities of output from the land. Fewer farm owners and farm workers have meant fewer farm families, as the victims of farm consolidation have been pushed from rural areas. Moreover, farm families have tended to become smaller in size, requiring fewer teachers or other local business and service workers in shrinking hamlets. Should the rural settlement structure of the Midlands eventually show evidence of stability, then it may become known just how few farmers must remain to operate the long-striding giant machines that have come to be used to plough, plant, cultivate and harvest the bountiful land of the region. If computer vision is ever perfected, those self-propelled long-striding giant machines may become self-guided as well, perhaps threatening to leave the Midlands as empty of humans as a robotized factory.

References

Agnew, J. (1987) *The United States in the World Economy*. Cambridge University Press, New York.

Associated Press (1990) U S farm population drops again. *Lincoln Journal-Star* 8 November, p. 19.

Brown, L. (ed.) (1985) *Audubon Society Nature Guides: Grasslands*. Knopf, New York.

Dauel, J. and Beals, J. (1988) Majority of students favor initiative 300. *Daily Nebraskan* 87, No. 108, pp. 1 and 3.

Dyson, L. (1989) History of federal drought relief programs. *Rural Development Perspectives* 6, 6–7.

Federal Writer's Project (1939) *Nebraska: A Guide to the Cornhusker State.* Reprinted 1979, University of Nebraska Press, Lincoln, Nebraska.

Garrett, W.E. (ed.) (1988) *Historical Atlas of the United States.* National Geographic Society, Washington DC.

Hart, J.F. (1986) Change in the Corn Belt. *Geographical Review* 76, 51–72.

Hart, J.F. (1987) The persistence of family farming areas. *Journal of Geography* 86, 198–203.

Ivey, J. (1991) 100 years after heyday, Cass County grows again. *Omaha World-Herald* 24 March, pp. B1 and B8.

Jones, E. (1968) *The Plains States; Time-Life Library of America.* Time-Life Books, New York.

Jordon, S. (1988). New equipment awaits farmers' bids. *Sunday World-Herald* 21 February, pp. 1-M and 7-M.

Lee, S.P. and Passell, P. (1979) *A New Economic View of American History.* Norton, New York.

McCarty, H.H. (1940) *The Geographic Basis of American Economic Life.* Harper and Row, New York.

Porter, J (1989) Droughts influence settlement patterns, both yesterday and today. *Rural Development Perspectives* 6, 2–7.

Rasmussen, W.D. (1982) The mechanization of agriculture. *Scientific American* 247, 48–61.

Reeder, R.J. and Glasgow, N.L. (1990) Nonmetro retirement counties' strengths and weaknesses. *Rural Development Perspectives* 6, 12–17.

Roseman, C.C. (1985) The population of the Midwest: changing composition and distribution. In Checkoway, B. and Patton, C.V. (eds), *The Metropolitan Midwest.* University of Illinois Press, Urbana, Illinois, pp. 29–55.

Semple, E.C. (1903) *American History and Its Geographic Conditions.* Houghton Mifflin, Boston.

US Bureau of the Census (various dates) *Statistical Abstract of the United States.* US Government Printing Office, Washington DC.

US Bureau of the Census (1931) *Fifteenth Census of the United States: 1930; Volume I Population, Number and Distribution of Inhabitants.* US Government Printing Office, Washington DC.

US Bureau of the Census (1960) *Historical Statistics of the United States, Colonial Times to 1957.* US Government Printing Office, Washington DC.

US Bureau of the Census (1983) *1980 Census of Population, Volume I, Characteristics of the Population.* US Government Printing Office, Washington DC.

US Bureau of the Census (1983) *County and City Data Book, 1983.* US Government Printing Office, Washington DC.

US Bureau of the Census (1986) *Local Population Estimates. Current Population Reports, Series P–26, No. 85–52-C.* US Government Printing Office, Washington DC.

US Bureau of the Census (1987) *Farm Population, Current Population Reports, Series P–27, No. 60.* US Government Printing Office, Washington DC.

US Bureau of the Census (1988) *County and City Data Book, 1988 on diskettes [Machine-readable data]*. US Bureau of the Census, Washington DC.

US Bureau of the Census (1989) *1987 Census of Agriculture Volume 1, Geographic Area Series, Part 51, United States Summary and State Data*. US Government Printing Office, Washington DC.

US Bureau of the Census (1990a) *State Population and Household Estimates: July 1, 1989, Current Population Reports, Population Estimates and Projections, Series P–25, No. 1058*. US Government Printing Office, Washington DC.

US Bureau of the Census (1990b) *Census of Agriculture 1987 on CD-ROM [Machine-readable data]* US Government Printing Office, Washington DC.

US Bureau of the Census (1990c) *Residents of Farms and Rural Areas: 1989; Current Population Reports, Population Characteristics, Series P–20, No. 446*. US Government Printing Office, Washington, DC.

US Bureau of the Census (1991a) *Census of Population and Housing, 1990: Public Law (P.L.) 94–171 Data on CD-ROM [Machine-readable data]*. US Bureau of the Census, Washington DC.

US Bureau of the Census (1991b) *Census and You*. Vol. 26, No. 4, US Government Printing Office, Washington DC.

Wardwell J.M. and Brown D.L. (1980) Population redistribution in the United States during the 1970s. In: Brown, D.L. and Wardwell, J.M. (eds) *New Directions in Urban–Rural Migration: The Population Turnaround in Rural America*. Academic Press, New York, pp. 5–36.

Wilkinson, L. (1990) *SYSTAT: The System for Statistics*. SYSTAT Inc., Evanston.

5

THE POLITICS OF PARTNERSHIPS: FARM WOMEN AND FARM LAND, ONTARIO

Fiona Mackenzie

It's like coming into this farm – you see this nice little house and the pillars, you know – very, very deceiving. It's just like the outside: we freeze to death in this place (in the winter) and in the summer we die of heat upstairs. It's very, very deceiving. On the outside it looks very nice, but inside it's totally different. So farming is the same way . . . there is a lot of clash, constant clashing.

(Member of Women for the Survival of Agriculture, May 1989)

In a context both of a deepening crisis in Canadian agriculture and of an increasing politicization of women's insecurity in the rural environment in the 1980s, the objective of this chapter is to explore the ways in which farm women in Eastern Ontario have organized around the issue of restructuring the family farm. Specifically, the chapter examines how members of Women for the Survival of Agriculture (WSA), a network of farmers in Stormont, Dundas and Glengarry counties, which emerged as a strong political voice in the 1970s and 1980s, centred their strategy on the creation of an oppositional or 'reverse' discourse (Foucault, 1978, p. 101) around farm partnerships. The argument shows that the effectiveness of their action is based on the strategic manipulation of two contradictory ideologies. On the one hand, WSA subscribes to a very powerful nationalist ideology wherein the family farm, through the provision of basic food supplies, becomes a symbol for the survival of Canadian sovereignty in the face of farm debt, the Canada–US Trade Agreement (CUSTA), and negotiations ongoing in the Uruguay Round of GATT. On the other, through an engagement in explicitly feminist politics, WSA challenges the hierarchical relationships (the 'farmer' and 'his wife') implied in the mythology surrounding the family farm. Discursively, members of WSA frequently disguise or overlay a feminist agenda – here of equality in farm partnerships – whose objective is the deconstruction of male hegemony in rural life, with a language that espouses the ethic of the family farm and rural security. It can be suggested that the family farm becomes 'a symbolic framework' (Chanock, 1989, p. 81) within which gendered dimensions of power are negotiated.

In conceptual terms, this study draws on the fertile ground unearthed as postmodernists expose the essentialism and foundationalism of much feminist analysis, which engages in binary reductionism (the either/or of domination/subordination or material/ideological) (Stamp, 1989; Fraser and Nicholson, 1990), and feminist scholars counter with an insistence on a political agenda that recognizes the connections within what Eisenstein (1988, p. 10) refers to as 'a radical pluralist epistemology of power relations' and which is missing in postmodernism (Diamond and Quinby, 1988; Fraser and Nicholson, 1990; Hartsock, 1990). Those who, like Diamond and Quinby (1988), Eisenstein (1988), Martin (1988), Sawicki (1988) and Fraser and Nicholson (1990), argue the case for complementarity between the two scholarships, recognizing that the objective of both is to deconstruct previously unrecognized modes of domination (Diamond and Quinby 1988), nevertheless point to the need for feminist analysis to explore the relationships among discourses and to question why certain discourses achieve hegemony.

Discourse here may be understood succinctly as the struggle, or negotiation, over meaning. To follow Foucault (1978), discourse is the site wherein relations of power and knowledge are articulated, where, as Turkel (1990) points out, procedures of inclusion or exclusion are defined. It may be, in Foucault's words (1978, p. 101), 'both an instrument and an effect of power, but also a hindrance, a stumbling block, a point of resistance and a starting point for an opposing stratgegy.' For Eisenstein, an analysis of discourses centres on

> the politics of language and knowledge – the awareness that power
> is constructed in and through language. . . . Discourse focuses on
> the importance of context within meaning and the open-texturedness
> of reality.
>
> (Eisenstein 1988, pp. 10–11)

In so far as feminism is a 'reverse discourse' (Foucault, 1978, p. 101) that struggles to prevail against theories of male hegemony (Sawicki, 1988), meta-theory and meta-narratives (for example, Fraser and Nicholson, 1990), it proceeds in part through genealogical research (for example, Foucault, 1980, pp. 83–84) to reconstruct those discourses that have been 'subjugated' or silenced historically, thereby giving visibility to women's modes of resistance and creating the means through which a new politics may emerge (Ferguson, 1984, p. 155).

With reference to the issue of farm partnerships, this chapter demonstrates how members of WSA have used to advantage what Martin (1988, p. 13) refers to as a position of 'internal exclusion', their particular claims to power and knowledge, in the creation of a language of resistance. The second part of the chapter sketches the legal context against which WSA's actions may be placed, before examining the action taken by WSA in promoting farm partnerships and, finally, the broader context within which the policies surrounding them may be understood.

Legal Spaces

The uproar that greeted the Supreme Court of Canada's decision in 1975 in the *Murdoch v. Murdoch* case (Canada, Supreme Court, 1975) sparked not only widespread publicity of the inequities farm women face vis-à-vis the land but, through the reasoning of the dissenting judge, Bora Laskin, paved the way for subsequent legal change.

In *Murdoch v. Murdoch*, Mrs Murdoch claimed entitlement to one-half interest in the land and other property used in the joint operation of the farm in a divorce settlement. Despite evidence that she had been an active participant in the operation of the farm for over 20 years, having full responsibility for the running of the farm during her husband's frequent absences, the trial judge found that she had done only what 'any ranch wife' would do, denying her a share in the property. The Supreme Court concurred, arguing on the basis of the doctrine of 'resulting trust' that such work did not create an interest in the land, i.e. there was no proof of intent to share property registered in one name. In his dissenting judgement, Judge Laskin drew on the principle of 'unjust enrichment' found in the doctrine of 'constructive trust'. This principle allows that a person who has acquired legal title to property, but who may not retain such interest in good conscience, becomes a trustee of that interest.

Public and, particularly, women's response to the Supreme Court's ruling had the effect of opening the way for two landmark cases. *Rathwell v. Rathwell* (Canada, Supreme Court, 1978) and *Pettkus v. Bekker* (Canada, Supreme Court, 1980) illustrated that, increasingly, women's economic contribution to the farm, whether through management of the home or through field or barn work, was legally recognized in the event of division of assets in the dissolution of a marriage. In the case of *Rathwell v. Rathwell*, title to all three farm properties was registered in the husband's name only, but equity used in the down payment for the purchase of each property came from a joint account to which both husband and wife had contributed equally. In this case, where the wife had worked both in the house and on the farm, the court found that both a 'resulting trust' and a 'constructive trust' were present. The decision recognized that it was only through the wife's work that the husband was able to acquire the land.

In the case of *Pettkus v. Bekker*, in 1980, the use of 'constructive trust' as a basis for judgements in the case of divorce was extended to common-law partners. During the first 5 of the 20 years that Rosa Bekker and Lothar Pettkus had lived together, Bekker had paid all expenses so that Pettkus could save to purchase a farm. Through her financial and labour contribution, they built and maintained a profitable bee-keeping business. The Supreme Court ordered an equal division of the assets, but not of the property, on termination of the relationship. Tragically for Bekker, a further 6 years were spent in trying to enforce a decision that had itself taken 6 years to reach. Pettkus used every means at his disposal to delay division of the assets and, in despair as a final attempt to change the legal system, Bekker took her own life in November 1986.

Pettkus v. Bekker illustrates very clearly the insecurity farm women face where rights to the value of assets are recognized rather than an interest in property. In the 1985 case of *Sorochan v. Sorochan* (Canada, Supreme Court, 1985), the Supreme Court set a legal precedent in recognizing property rights. The issue here was whether the doctrine of 'constructive trust' could be used to claim rights for the female partner when the land was owned by the male partner before marriage.

In this case, Mary Sorochan, who had lived (unmarried) and worked with her partner for 42 years, was responsible for all the labour in the house, including raising six children, and much of the farm labour. Her partner was frequently absent as a travelling salesperson. When they began living together, the six one-quarter sections of land were jointly owned by her partner and his brother, but were later divided between them. The Supreme Court upheld the trial judge's decision ordering that one of the three sections be transferred to her with additional financial compensation. On the grounds that the requirements for 'constructive trust' were met, particularly that her work conferred a benefit on her partner, the Supreme Court thus reversed the Court of Appeal's ruling that had stated that there was no necessary connection between property rights and labour to form a 'constructive trust.' Importantly, the Supreme Court's decision considered that contribution to the purchase price of property was not necessary to establish a property interest. Maintenance of the value of the land was sufficient. Further, in the decision, Mary Sorochan's household labour was deemed of economic value.

In Ontario, the *Family Law Act* (Ontario, 1986) gives each spouse a right to half the net value of the farm (Section 5[7]) but, despite an expanded definition of 'family assets' or 'family property,' does not provide for an equal division of property. Under this Act, marriage property includes any gain made during the marriage of property by one of the spouses. Tangible assets, such as land, house, furniture, farm machinery and car, are included, as well as stocks, bonds, insurance policies and any income from property belonging to one of the spouses before marriage, and any increase in the value of such property or inherited property during the marriage (Meanwell and Glover, 1985, with 1986 addendum; Boivin,

1987). As Boivin (1987) points out, a farm acquired during marriage would be subject to equal division, but the farm itself would remain the property of the spouse with title. Only in isolated cases, as in *Mackay v. Mackay* (Ontario District Court, 1986) has a wife been awarded a proprietary interest under the *Family Law Act* 1986.

But there is a further issue that arises for farm women who are not protected by a legal partnership agreement. As Hendin (1989, p. 66), a lawyer who has worked with WSA, points out, interests secured under the Ontario *Family Law Act* 1986 'kick in' only on dissolution of the marriage. The case of *Atkinson v. Atkinson* and *Lloyds Bank v. Atkinson and Atkinson*, heard before the Supreme Court of Ontario in 1988, illustrates the point. When her marriage broke down, Ruth Atkinson claimed 50% of the assets of the farm on which she had worked with her husband, and priority in her claim over that of Lloyds Bank which had loaned money to her husband and which had subsequently been lost through speculation. In this instance, Gordon Atkinson of Simcoe, Ontario, the owner of a dairy enterprise (valued at C$2.5 million), which bred Holsteins for sale internationally, had used all his business interests, including milk quota, as collateral for a loan of several million dollars. The Bank negotiated with the husband, treating him as the sole owner (interview, WSA member, July 1991; Hendin, 1989). In his ruling, Judge Thompson found that the Bank had not been negligent in negotiating the loan solely with the husband. In his words:

> Everything connected with the farming and livestock operations was in the name of Mr Atkinson, as were the milk cheques. Everything indicated that the husband was the only person carrying on the business and in my view the Bank was entitled to treat Mr Atkinson as such. . . . I should add that there was no evidence adduced to support the position that Mrs Atkinson was a partner with her husband in the business.
>
> (Ontario, Supreme Court, 1988, *Reasons for Judgement*, pp. 11, 13).

For woman farmers, the point as expressed by Middleton is that, without a legal partnership, assets may wither away before the dissolution of a marriage without the spouse having any say in the decisions (interview, July 1991).

Partnerships and Politics

Against this background, and the fact that only 37% of partnerships in Eastern Ontario were legally registered with the Provincial Ministry of Consumer and Commercial Relations (Watkins, 1985, p. 4), a major objective of the WSA leadership during the 1980s has been to politicize its members, and farm women more generally, about farm partnerships. This section of the chapter argues that WSA has been careful to construct a language around farm partnerships that, although containing an explicitly feminist agenda, is cloaked in an economic guise. The final section of the chapter relates this issue more broadly to WSA's actions.

The economic rationale for a legal partnership is straightforward, and prominent in WSA's booklet: *Cover Your Assets: A Guide to Farm Partnerships* (1987). First, significantly, on sale of the farm, each partner is eligible for a C$500 000 lifetime capital gains exemption. Under the terms of the capital gains exemption introduced by the federal government subsequent to effective WSA lobbying in 1985, each partner may claim full exemption. Previously, the farm had been treated as one undifferentiated spousal unit (interview, WSA member, July 1991). Second, there are frequently tax advantages to 'income splitting' (WSA, 1987, p. 2). Third, a legal partnership agreement allows two partners to apply for a maximum loan of C$600 000 from the Farm Credit Corporation (FCC). The maximum loan for a family farm with one qualifying applicant is C$350 000 (Canada, *Farm Credit Act*, 1985). Fourth, in the event of the husband's death, where a legal partnership is in operation, the previous establishment of the wife's credit rating will allow the family farm to continue to function without undue stress (WSA, 1987, p. 2).

But through the symbolism of the family farm, an altogether different, and at times contradictory, feminist agenda is smuggled in. This agenda insists on ensuring farm women's security, based, as Judy Hendin argued so forcefully at the Fourth National Farm Women's Conference, on their self-reliance (Hendin, 1989). Fundamentally, this message challenges farm women to distinguish between their marriage and business partnership. The argument proceeds as follows:

> When we marry we assume that our relationships will last forever;
> or that if difficulties arise, we will have cared enough for each
> other that we will want to and, therefore, will ensure that all family
> members are well looked after. However, this picture of economic
> caring within the relationship does not always occur. Statistics tell
> us that many women, in particular, live in poor economic conditions
> after marriage breakdown.
>
> (WSA, 1987, p. 3)

A business partnership becomes 'preventive medicine' (WSA, 1987, p. 3), as a legal arrangement ensures that the time, money and energy of each partner in the operation of the farm enterprise is documented:

> . . . it is a safeguard to establish a more formal legal relationship
> between those members of a family involved in a farm business. It

provides a spirit of trust and mutual respect based on an orderly approach to the risks of life, both business and personal.

(WSA, 1987, p. 3)

Shared ownership of the business allows each partner to increase security through contributions to a Registered Retirement Savings Plan and the Canada Pension Plan.

The resistance with which the notion of farm partnerships has been greeted illustrates, however, that the issue fundamentally disputes the hierarchical and patriarchal lines of authority within the family farm and its presumption of 'free' labour of household members. One WSA member spoke of the threat men frequently feel when faced with a decision to legalize an informal partnership:

> If the man is the sole owner and says 'You're a partner dear' – what advantage is there for him to put her name on. A mental thing happens to the guy – they see themselves in a power position. With someone else's name, they see their power/decision-making threatened . . . When men lose physical ownership, they feel they lose their authority and power. If the man is the sole proprietor, he is dominant. Is it likely that a dominant psyche is going to reduce their power?

The case of one WSA member may serve to indicate this resistance, in this case stemming from the father-in-law and mother-in-law rather than the husband. Extracts from an interview with K.M., in May 1989, illustrate her perceptions of what led up to her purchase of her father-in-law's interest in the farm on which she and her husband lived. Her husband had previously bought 50% interest in the farm.

> K.M. I don't agree with it that the farmer owns everything and the wife works on it and there isn't any money to pay her and she seems to be content, but I guess she thinks this is the way everybody else does it. I couldn't agree with that because I was an independent person all my life. So I wanted to make sure I wouldn't get caught in that situation. So I said, I know there is not going to be any money to pay me for all the work I'm going to do, therefore I want ownership of it.

> F.M. How was [the purchase] viewed by the rest of the family?

> K.M. There was chaos. War. I fought with my father-in-law constantly because I always kept my ground and . . . in the past you didn't do that and so he wasn't used to that you know.

> F.M. Was your mother-in-law behind you in any of this?

K.M. No, she wasn't. I think there was jealousy that I was like she
never worked on the farm. There was a lot of jealousy that
I was helping my husband and doing the milking and all that
stuff so there was a clash there too between my mother-in-
law and myself. And that happens in just about every family
but you don't see that.

K.M. had no trouble in obtaining a loan from the Farm Credit Corporation (FCC) to purchase her father-in-law's 50% interest in the farm,
but she did run into problems when she applied to the Ontario Ministry
of Agriculture and Food to reduce her interest costs from 13.5% to 9.25%
under the Beginning Farmer Assistance Program (BFAP) in 1984. She
qualified for the BFAP in all respects except that she was the spouse of
an existing farmer. Under the BFAP, a husband/wife partnership (she and
her husband have a 50–50 legal partnership), unlike a father/son or
brother/brother partnership, is defined in the context of a 'spousal unit'
as equivalent to one person. Angered that denial of assistance was based
on her marital status rather than her business partnership, her subsequent
politicization of the issue has led to the case being reviewed by the Ontario
Human Rights Commission. Ironically, K.M. did obtain a rebate on
interest through her husband's application under the Ontario Interest
Rebate Program – he had had to co-sign K.M.'s loan from the FCC. In
her words, 'It was my debt, but he got the interest rebate on his debt and
my debt.'

The publicity around the case led to a change in BFAP language,
allowing a spouse to claim an interest rebate when another farm (i.e. not
the one on which the applicant is living) is purchased; but, K.M. argues,
'The new language just masks the discrimination, as few women would
have the financial or other resources to purchase a separate farm.'

Partnerships in Context

Farm partnerships are one instrument used by WSA to constuct an alternative discourse. An explicitly feminist objective is to deconstruct thereby
an image of farm women as 'farm wives', 'invisible pitchforks' or 'gofers',
and to construct an alternative image of women as business partners.
Through newsletters (the *WSA Newsletter* is published bimonthly), commissioned studies have questioned how women's work has been counted
(for example, Watkins, 1985), lobbying with Members of the House of
Commons and on Standing Committees (for instance, with respect to the
capital gains exemption and in order to ensure that Statistics Canada
allows more than one operator to be listed per farm). Also through
extension courses such as *Survival Techniques for Farm Women* run annu-

ally since 1980 at Kemptville College of Agricultural Technology, WSA has evolved an effective strategy to challenge and restructure power relations in agricultural practice – from within the household, to provincial and federal arenas.

Intertwined with a feminist politics is a discourse of resistance against the demise of the family farm in the face of an agricultural crisis. For, although the majority of farmers in Stormont, Glengarry and Dundas counties are dairy farmers, and protected through supply management (the Ontario Milk Marketing Board) from the volatility and present depression of commodity prices on the world market, high interest rates and spiralling production costs have affected all farmers, and particularly those of a generation encouraged to expand and capitalize their operations to qualify for bank or Farm Credit loans. In this economic climate, WSA has, through lobbying, public speaking, workshops and newsletters, mobilized around the symbolic importance of the family farm for the survival of Canadian agriculture. It has fought for parity, for the maintenance of supply management in the context of CUSTA and GATT negotiations, for relief from farm debt and, most recently, around the issue of 'pure milk'.

The WSA leadership is aware of the apparent contradictions within the discourse it has created around the family farm. In the words of one former member of the executive: 'We had to use the family farm or we would not have got off the ground . . . we would have turned the women off.' Or, in the words of a second member: 'As a new movement, this is one of the biggest barriers you face – you fight not to alienate . . . We had to take a farm women's issue and make it an agricultural issue to get by . . . [to speak with] a forked tongue.' The second member continued, 'Our goals were not contradictory to us, but we had to overlay our goals with a contradiction' (interviews, December 1991).

In essence, the discourse of farm partnerships created by WSA challenges male hegemony in agricultural practice. It interweaves an economic and feminist struggle from the control of labour and land at the household level with a discourse that resonates in the world of banks, courts of law, and federal and provincial policy. This strategy directly assaults the blurring of unequal relationships that have always been part of the family farm, premised, as Rankin (1987) argues, on the appropriation of frequently unpaid labour of an adult woman and children. Members of WSA have drawn from their particular vantage point of 'internal exclusion' (Martin, 1988, p. 13) to conjure to their advantage, as symbol, an ethic of the family farm and rural stability. As Scott (1985) explains in a very different context, the struggle is not only a struggle over work and property rights: 'It is also a struggle over the appropriation of symbols, a struggle over how the past and present shall be understood and labelled' (1985, p. xvii). WSA has thus created a discourse that links their particular claims to power/knowledge – their 'localized resistance' (Foucault, 1978, p. 96) – to other points of struggle, which rest on the common concern of agricultural producers on family farms to survive. The degree to which the organization has successfully managed to achieve its objectives appears to rest on the extent to which the symbolism of the family farm has been created as part of the discourse.

Acknowledgement

Funding for this research was provided by the Social Sciences and Humanities Research Council of Canada.

References

Boivin, M. (1987) Farm women: obtaining legal and economic recognition of their work. In: Canadian Advisory Council on the Status of Women, *Growing Strong: Women in Agriculture*. CACSW, Ottawa, pp. 49–90.

Canada, House of Commons (1985) Minutes of Proceedings and Evidence of the Standing Committee on Finance, Trade and Economic Affairs. Chair: Don Blenkarn. Issue No. 17, Thursday, 14 March 1985.

Canada, Supreme Court (1975) *Murdoch v. Murdoch* (1975, 1 S.C.R. 423).

Canada, Supreme Court (1978) *Rathwell v. Rathwell* (1978) 2 S.C.R. 436, 83 D.L.R. (3d) 289.

Canada, Supreme Court (1980) *Pettkus v. Bekker* (1980) 2 S.C.R. 834, 117 D.L.R. (3d) 257.

Canada, Supreme Court (1985) *Sorochan v. Sorochan* (1985) 2 R.F.L. (3d) 225.

Chanock, M. (1989) Neither customary nor legal: African customary law in an era of family law reform. *International Journal of Law and the Family* 3, 72–88.

Diamond, I. and Quinby, L. (eds) (1988) *Feminism and Foucault: Reflections on Resistance*. Northeastern University Press, Boston, pp. ix–xx.

Eisenstein, Z.R. (1988) *The Female Body and the Law*. University of California Press, Berkeley.

Ferguson, K.E. (1984) *The Feminist Case Against Bureaucracy*. Temple University Press, Philadelphia.

Foucault, M. (1978) *The History of Sexuality. Volume 1*. Vintage Books, New York.

Foucault, Michel (1980) Two lectures. In: Gordon, C. (ed.) *Power/Knowledge: Selected Interviews and Other Writings 1972–1977 by Michel Foucault*, Pantheon Books, New York, pp. 78–108.

Fraser, N. and Nicholson, L.J. (1990) Social criticism without philosophy: an encounter between feminism and postmodernism. In: Nicholson, L.J. (ed.) *Feminism/Postmodernism*, Routledge, New York, pp. 19–38.

Hartsock, N. (1990) Foucault on power: A Theory for women? In: Nicholson, L.J. (ed.) *Feminism/Postmodernism*. Routledge, New York, pp. 157–75.

Hendin, J. (1989) Speech published in *The Future of Farm Women and Agriculture*. Proceedings of the Fourth National Farm Women's Conference, 16–18 November, Saint John, New Brunswick.

Martin, B. (1988) Feminism, criticism, and Foucault. In: Diamond, I. and Quinby, L. (eds), *Feminism and Foucault: Reflections on Resistance*. Northeastern University Press, Boston, pp. 3–19.

Meanwell, C. and Glover, G. (1985) *To Have and To Hold: A Guide to Property and Credit Law for Farm Families in Ontario*. Concerned Farm Women (with 1986 addendum), Chesley, Ontario.

Ontario (1986) *Ontario Family Law Act*, S.O. 1986, C.4.

Ontario District Court (1986) *Mackay v. Mackay* Walkerton, Doc. No. 48/85, Mackay J., May 26.

Ontario Supreme Court (1988) *Atkinson v. Atkinson and Lloyds Bank v. Atkinson and Atkinson* No. 958/86.

Rankin, L.P. (1987) Beyond the kitchen and the cornfield: the political activism of Ontario farm women. Unpublished MA Thesis, Carleton University, Ottawa.

Sawicki, J. (1988) Feminism and the power of Foucaldian discourse. In: Arac, J. (ed.), *After Foucault: Humanistic Knowledge, Postmodern Challenges*. Rutgers University Press, New Brunswick, pp. 161–78.

Scott, J. (1985) *Weapons of the Weak: Everyday Forms of Peasant Resistance*. Yale University Press, New Haven and London.

Stamp, P. (1989) *Technology, Gender and Power in Africa*. International Development Research Centre, IDRC-TS63e, Ottawa.

Turkel, G. (1990) Michel Foucault: law, power, and knowledge. *Journal of Law and Society* 17, 170–93.

Watkins, S. (1985) *What are You Worth? A Study of the Economic Contribution of Eastern Ontario Farm Women to the Family Farm Enterprise*. Women for the Survival of Agriculture, Winchester, Ottawa.

Women for the Survival of Agriculture (1975–1991) *Newsletter*. WSA, Winchester, Ottawa.

Women for the Survival of Agriculture (1987) *Cover your Assets: A Guide to Farm Partnerships*. WSA, Winchester, Ottawa.

6

RURAL AGEING: PERSPECTIVES FROM THE US AND UK

Glenda Laws and Sarah Harper

The greying of rural areas is a topic that has attracted interest in recent years, both in the UK and US. The 1986 Conference of the Gerontological Society of America devoted a special session to 'Ageing in Rural America', while the British Society of Gerontology and the British Sociological Association held a joint one-day seminar in 1988 entitled 'Rural Ageing'. It is worth noting, however, that the latter seminar attracted more interest from social service departments than from British academics! Indeed the quantity of American research into or associated with ageing in rural areas far outstrips that produced by British researchers. The several bibliographies and texts on ageing in rural US (Krout, 1983, 1986; Coward and Lee, 1985), are matched in the UK by Wenger's work on elderly networks in rural Wales (Wenger, 1984, 1991), though there is a small but growing collection of papers specifically on the rural aged (Mosely, 1978; Victor et al., 1983; Jones et al., 1985; Harper, 1987; Gant and Smith, 1988, 1991). Indeed, in the US the desire to understand the causes and consequences of the ageing of the rural population sparked the growth of a subdiscipline, rural gerontology, in the early 1970s. In part, this was a product of some policy initiatives on the part of the federal government. Changes to the *Older Americans Act* in the 1970s recognized the plight of the rural elderly and mandated provision of certain services. Academic data collection and analysis grew as the infant programmes were evaluated. As our knowledge of the living conditions of the rural old expanded, so too did the pressure for further policy initiatives. Thus, there has been a reciprocal relationship operating between the policies aimed at the rural elderly and the types of research being conducted. Two points clearly emerge from a close reading of the rural gerontology literature. First is the relatively small input by American geographers, limited to a few topics, most notably residential environments and migration/mobility studies. The second point to be noted is the paucity of work that really tackles some of the conceptual issues surrounding the notion of a rural gerontology. Almost all these papers deal with the conceptual issues that frame their particular question: for example, migration theory, residential preferences models etc. However, very few ask 'what is rural about rural ageing' (Rowles, 1988), and thus do not make it clear why we should be looking at the rural experience of ageing, or the rural elderly, as a distinct category.

Krout (1988) suggests that rural gerontologists require a concentrated effort to understand the life circumstances and experiences of the rural elderly and to explore if, how, why, and under what conditions the salient characteristics of rural environments are influential. It is thus important to identify aspects of the rural environment that interact with the process of ageing, and examine how these shape the networks and systems that affect elderly people. As Rowles (1988) suggests, in order to understand these interrelationships, it is necessary to combine a consideration of the macro-ecological processes with an examination of the ecological, sociocultural and phenomenological contexts of those who are growing old in rural environments.

This chapter reviews the literature on the ageing of the rural population in the United States and the United Kingdom. It shows that the growth of the older-age cohorts in rural counties is occurring not only because of the natural ageing of an in-place population, but also, in a few selected destinations, because of the immigration of retirees from non-rural places. This ageing of the rural population presents special planning problems because of the difficulties of service provision in sparsely populated areas. We focus on the empirical 'reality' of a changing rural population, pointing to: (i) some of the major findings regarding the well-being of the older, and ageing, population, and (ii) the links between demographic change and economic restructuring. Rural areas are often associated with certain ecological, occupational and sociocultural attributes (Bealer *et al.*, 1965). To many, these characteristics collectively conjure up images of agricultural regions dotted with farmhouses and small service centres. Rural areas are, however, diversifying, and the service and manufacturing components are becoming larger as the number of people engaged in primary activities declines. How does this impact on the demographic make-up of the rural place?

Rural Ageing: A Geographical Challenge and a Challenge to Geography

Ageing in place provides one of the most important challenges to social planners in the coming decades if for no other reason than the rapidity with which the process is occurring. There is, in advanced industrial nations, an unprecedented growth in the number and proportion of people labelled as 'old'. Such people require support services at the very time when most of them find their income greatly reduced. Ageing in rural areas is particularly problematic because of the 'inefficiencies' involved in delivering a small number of service units over a relatively large spatial jurisdiction. Ageing offers, then, an immediate planning and political challenge that has a very real geographical dimension. It must be admitted, however, that, given the importance of the ageing population, the degree of geographic attention to the question of the rural elderly in the US and UK is limited, despite its quality. Further, when geographers do attend to these problems, they often publish outside the discipline's journals. This is a positive sign in that their work is being read by a broader population of social gerontologists, a group that prides itself on the applied nature of its work. However, the negative side of this is that within the discipline, questions concerning the rural elderly are marginalized and not seen as part of the mainstream of geography's current agenda. There is a challenge to the discipline then, to become involved in constructive debates about how best to serve our old citizens as well as debates about the intellectual merits of such work.

The current demographic structure

In 1980, 11.3% of the total population of the United States was 65 years or older. The proportion in rural areas varied: just under 11% in total, but as high as 15.4% in rural areas of between 1000 and 2500 people. The total non-metropolitan figure stood at 13%. Almost 7.5 million 'old people' lived in non-metropolitan areas. As might be expected, the proportions of elderly people (as a percentage of the total population) living outside metro areas were highest in the Northeast (13.2%) and North Central (17.0%), and lowest in the West (13.2%) and South (12.9%). In the North East and the West, 75% of the elderly live in metropolitan urban areas; in the North Central region the figure is 65%; and in the South it falls to 53%. That is, almost half the old people in the South live in either a non-metropolitan or a rural environment. More than 30% are found in rural places. Figures for 1990 show that whereas 21.7% of the total US population lived in non-metro areas, 26.2% of people over 65 years did. This figure rises to over 28% for those over 85 years (McLaughlin and Jensen, 1991).

The percentage of Britain's population over 65 years stands at around 15%, one of the highest in the world, with the US at 11.3% and Canada at 11.4% (OECD, 1988). In addition the *rate of increase* in the older age groups is far higher than the average for Britain's population. Thus, between 1971 and 1981, the percentage increase among those of pensionable age (60/65) was 14.5%, and for those over 75, 27.7%, compared with 5% for the population as a whole.

In 1981, the distribution of the elderly population of England and

Wales was concentrated in four main bands running through the South West, which as a region averaged 20.7%, East Anglia (18.7%), Wales (18.4%), and the South East (17.8%), and in clusters in the rural North. In general these bands, which registered over 21% of their population as pensionable (60/65), were rural and typically coastal.

Detailed work on the 1981 census data by Warnes and Law (1984) indicated a subtle shift in elderly relocation patterns during the 1970s. While the immediate post-war period saw a concentration of elderly people into selected areas of the country, in particular the South, the 1971–1981 decade revealed a preference for suburban and semi-rural areas peripheral to traditional retirement zones. In addition, East Anglia and selected northern areas, especially southern Yorkshire and the Lake District, joined the category of key retirement areas. An analysis by Cross (1990) of percentage change by population group between 1971 and 1981, using the OPCS classification of district class, similarly confirms the growth of elderly populations outside traditional retirement zones, but still within rural and semi-rural districts.

Such dispersal is clearly not only related to growing affluence and thus car ownership among subsequent cohorts of elderly, as suggested by Warnes and Law (1984), but also to the internal demographics of these regions. *In situ* ageing, typically by only one of the original retirement couple, results in one person occupying a formerly two-person housing unit. Similarly, the spread of such retirement migration throughout the social classes is placing considerable demand on traditional areas, increasingly reflected in the market price of such property. Such decentralization may thus be a simple market response. The 1991 census will hopefully clarify these trends.

The reasons for concentrations of rural elderly

It became evident during the 1970s that a polarization of elderly people was occurring in most developed countries. Significant concentrations of elderly populations were emerging in three broad zones: the inner cities, rural areas and retirement-orientated areas (Logino *et al.*, 1984; Law and Warnes, 1976; Karn, 1977). In recent years these zones of attraction have been extended and a more dispersed pattern of location has developed (Grundy, 1987). Thus there is increasing evidence of elderly concentrations in the suburbs (Golant, 1987; Hugo, 1984); a decline of elderly populations in remote rural areas, as the *in situ* elderly population dies and, due to earlier out-migration of the younger population, is not replaced (Golant *et al.*, 1988); and the growth of residential concentrations outside the traditional retirement destinations, particularly in non-peripheral rural areas (Fuguitt and Tordello, 1980; Longino *et al.*, 1984; Hugo, 1984; Ormond, 1986; Rogers and Watkins, 1987; Harper, 1987).

This distribution is thus partly fuelled by the small, yet significant population who relocate their households in their later years (Warnes and Law, 1985; Cribier, 1980). There is, however, substantial variation in elderly migration rates between countries, in part linked to the rates of mobility demonstrated by the population as a whole. Thus within the highly mobile US, around one quarter of those over 60 years relocate their households in older life (Golant, 1987), compared with only 5% in the UK. Such inter-country variations also arise, however, for inherent differences within national data bases.

Geographers, along with other social gerontologists, have thus contributed to our understanding of how the current concentrations of the rural elderly emerged (Bohland and Rowles, 1988; Golant, 1987, 1990; Meyer and Speare, 1985; Meyer, 1987a, b; Graff and Wiseman, 1990; Watkins, 1990; Wiseman and Roseman, 1979; Gober and Zonn, 1983; Grundy, 1987). Two major processes have been operating and continue to reshape the geography of rural ageing. *In situ* ageing occurs in terms of absolute numbers as the local population grows older, and proportionally as the younger population leaves or is not reproduced. The second process is that involved in *the in-migration of older people* from non-rural communities or other rural centres. Both processes can result in an increase in the absolute numbers of the elderly (cohort growth) or an increase in the proportion of the elderly in any one place (increasing concentration).

An examination of the US situation exemplifies these processes. *In situ* ageing reflects, in part, the broader economic restructuring of the US urban and regional system. In 1920, 30% of the US population was classified as living in farm communities; only 2% of the total population was so classified in 1990. About 12% of this farm population in 1970 was aged 65 years or more; by 1987, this figure had increased to 14.1%, even though the absolute numbers had declined during that same period (US Bureau of the Census, 1989). According to the calculations of Bender *et al.* (1985), among the non-metropolitan counties of the United States, the highest proportions of old people are found in 'farming-dependent' communities where almost 16% of the total population is over 65 years. These are the counties that experienced population losses in the 1960s and only very small increases in the 1970s, when the rest of non-metropolitan US was experiencing a 'turn around' as the population growth there outstripped that of metropolitan areas (Fuguitt and Tordello, 1980). This increasing proportion of the elderly in the nation's farm communities can be accounted for in part by the current trend towards lower fertility rates; the higher fertility rates of the earlier decades of this century are overwhelming recent trends toward smaller family size, and thus the younger cohorts are not being reproduced at a rate that balances the ageing of the population born early this century. Another component of the explanation is the out-migration of younger people who search out education and employment opportunities in urban areas. At the local level, the impacts might be far more profound. For example, the 1980 US census showed that over 15% of the population in rural places of between 1000 and 2000 people were over 65 years of age. In general, there seems to be evidence to suggest that the smaller the population size of a place, the greater the proportion of older people (Clifford *et al.*, 1985). Smaller places, with fewer fiscal resources, might therefore have relatively high levels of demand. Further, small populations change more rapidly and thus the impact of ageing might be more marked (Watkins, 1990).

In contrast to those places that have aged due to the out-migration of younger cohorts, a significant number of counties have experienced a growth in the number and proportion of people over the age of 65 years, due to the in-migration of retirees. The in-migration of the elderly also reflects broader changes in the economy and attitudes toward living environments. Perceptions of the problems of rural living have caused some people to leave urban areas (Cebula, 1974; Williams and Sofranko, 1979; Sofranko *et al.*, 1981). Young people do not always have that choice because of the need for employment; but the residential search behaviour of the elderly is freed from the constraints of employment opportunities and so amenity features can become a higher priority (Serow, 1987; Speare and Meyer, 1988; Gober and Zonn, 1983). Many non-metropolitan areas have taken on the status of 'retirement communities'. Again, according to Bender *et al.* (1985), 515 counties accounted for 26% of the rural non-metropolitan population aged over 65 years in 1980. Unlike those counties described in the preceding paragraphs, these counties experienced high rates of population growth between 1960 and 1980, and are characterized by large financial inputs by way of transfer payments. Further, a large service sector is associated with this type of ageing of a community. These communities are not the ones that see young people leaving in search of jobs, since their economies tend to be relatively strong. Bender *et al.* write:

> Retirement counties usually have robust economies compared with counties where people retire without moving to another place. The labour force is likely to be growing because young people also move into destination retirement areas to work
>
> (Bender *et al.* 1985, p. 18)

Economic development planners are therefore keen to attract the relatively mobile, healthy and wealthy elderly who move on retirement. This is, in fact, a small proportion of the total old population. Longino (1990) states that in any 5-year period, only about 10% of people over 65 years will move across state or county boundaries. In another paper he estimates that of the 35.6 million people aged more than 60 years in 1980, 1.6 million, or 4.5%, had moved across state boundaries in the preceding 5 years (Longino *et al.*, 1984, also see Bryant and El-Attar, 1984). The majority of the elderly who do migrate tend to move to the same types of residential areas (for example, urban to urban) (Clifford *et al.*, 1985). Therefore, despite the quantity of research on the migrating elderly, they are only a very small proportion of the total elderly population. However, their absolute numbers are significant and they bring with them not only government transfer payments, but also years of savings which are spent in the local economy as they purchase goods and services to enhance their retirement living. Crown (1988) has estimated that Florida has gained 1 billion dollars because of the migration of the elderly from New York alone. Large concentrations of old people are, therefore, not necessarily a burden on the local economy.

But it must be remembered that the migrating elderly are a self-selecting group. Communities that do not receive such an affluent group of retirees might not enjoy such benefits. And so those communities that are ageing as a result of the loss of the younger labour force might suffer and find that it is in fact a burden to deal with the local ageing of the population. And it is yet to be seen what happens to the local economy and human services sector as the migrants age. Watkins (1990), for example, describes bed shortages in hospitals in resort areas in North Carolina as the population ages (also see Bryant and El-Attar, 1984).

The experiences

What kinds of experiences do the elderly have in rural areas? On almost any indicator of health, income, access to services etc., the rural elderly have been shown to be disadvantaged (Nelson, 1980; Steinhauer, 1980; Coward and Lee, 1985; Lassey and Lassey, 1985; Krout, 1986). There are also several studies that show that the rural elderly are comparable with their urban counterparts on a number of measures of well-being and accessibility to services (Meyer, 1981; Krout, 1984; Hiltner *et al.*, 1986). Often these conclusions are drawn from survey research or analysis of aggregated data which measure 'objective indicators'. The so-called 'rural-urban paradox' (Kivett, 1988), however, shows that regardless of those objective measures, which demonstrate that the living situation of the rural elderly is not so favourable as that of the urban elderly, the rural elderly are more likely to report satisfaction in their living arrangements than might be expected (Sofranko *et al.*, 1983). Efforts to understand this have come up with a number of possible explanations. Krout (1986) suggests that rural values place a premium on utility and it is, therefore, more likely that rural people will be satisfied with things so long as they are functional. The physical space of rural places also means that people are happier with minimal services or housing, so long as their sense of 'having roots' is not disturbed (Kivett, 1988). Finally, Kivett (1988) suggests that their 'level of relative deprivation' means that the rural elderly have fewer comparative objects and experiences and thus may have 'lower' expectations.

Unfortunately, relatively little of the work by geographers describes the experiences of the elderly living in rural areas. Graeme Rowles stands out as someone who has sought to move beyond official statistics to find out what it is like to be an old person in a relatively isolated rural place. His humanistic perspective centres 'on the evolving symbiotic relationship between the individual and an environmental context'. He argues that 'a person-centered focus facilitates deeper insight into the genesis of aggregate patterns of location and migration. It also facilitates more empathic appreciation of the underlying determinants of well-being and dilemmas of appropriate service delivery to the vulnerable elderly' (1986, p. 511). In his study of a small community of old people in rural Appalachia, Rowles describes the sense of insidedness experienced within a rural setting, the 'integration within the social fabric of the community', which allows for the accumulation of

> social credit derived from contributions to family and the community made over the span of each individual's life. Old people could draw upon a reservoir of accumulated social credit as they grew progressively more vulnerable and in need of assistance.
>
> (Rowles, 1983b, p. 302)

This social credit allows the functioning of an informal support network. Such a network is critical to the well-being of an elderly population that may not have access to a dense formal support network as might be available in an urban centre (Newhouse and McAuley, 1987). This informal network manifests itself in many ways including the 'surveillance zone' identified by Rowles (1981). Old people watch for each other from their windows or porches, visually monitoring each others' activities. Unusual breaks in a neighbour's activity pattern will be observed and can be investigated if there is no explanation for the break in routine. Similarly, frequent telephone calls help old people keep track of each other and unanswered calls might signal the need for some further investigation.

Rowles' investigation thus establishes the importance of informal (and internal) support in the ageing community. Despite the strength of these ties, some of the elderly people in his study found that they had to consider relocating as their health and mobility deteriorated (see also Aitian and Aitian, 1990). The prospect of moving is often resisted by old people, in both rural and urban areas. Rowles (1983a) describes the struggles as 'reconciling the physical, social and emotional support of a familiar environment with the desire to be close to family', and outlines the strategies of some people who resist the move for as long as possible by drawing upon the social credit they have accumulated.

In-depth work by Harper (1987), in the British context, has similarly emphasized the importance of informal support for the rural elderly. Her work highlighted the issue of residential mobility in our understanding of the rural experience of ageing, arguing that the reuniting of families in later life did not necessarily lead to close supportive relationships. Indeed spatial separation in early life established forms of interaction that did not change when the families moved near to each other as parents aged. This has important implications as post-war residential mobility in the UK is reuniting families, as well as separating them, and government policy is increasingly relying on these reunited families to care for ageing parents.

The Context: Restructuring Rural Regions

Two important points need to be noted when discussing the social recomposition of rural areas. First, demographic change is linked to economic restructuring. This is recognized most often in discussions that imply that demographic changes (for example, ageing, changing ethnic composition) have repercussions for the economy. So, we see articles on the 'economic and social implications of demographic change' (Meyer and Kern, 1990). A growth in the number and proportion of elderly people means that social and economic resources will have to be shifted away from younger age groups. But this is not a unidirectional relationship; the restructuring of the economy also has implications for demographic structure.

At an international level this is observable in changing migration policies. As certain national economies require more or less labour, regulations governing the number of working-age people permitted to enter are loosened and tightened. At a local level a similar, more informal, process is at work. As job opportunities leave an area so too do younger people seeking employment. Thus the local area ages proportionately. If new employment opportunities arise, young people might move into a region and thus the population becomes younger on average. Rural areas, in particular, have aged due to the out-migration of the working-age population. Places that experience this type of *in situ* ageing are, in effect, ageing by default.

But some rural areas are ageing by design. This raises a second point: rural restructuring and place ageing are not 'natural' processes over which policy-makers have little control. Local planners are actively encouraging the in-migration of the elderly because of perceived economic benefits (Dunbar, 1990). Not only do retirees bring wealth and assets, they also stimulate the local economy, especially in terms of their demands on the service sector. In early 1991, the Pennsylvania Department of Aging received a federal grant 'to initiate a project involving retiree-based economic development'. The general goals of the project were to: (i) promote a growth in retirement housing; (ii) create new commercial services; (iii) increase the production and consumption of services specifically targeted toward retirees; (iv) improve retailing practices; and (v) enhance employment of older workers. This is evidence of the proactive nature of some actions that promote the ageing of local places. As a means of accommodating the economic restructuring that surrounds them, local states are inviting old people, and the services on which they rely, to relocate. We have seen this with local 'bidding wars' to attract manufacturing industry; now we see it with the retirement industry.

Economic restructuring is indirectly responsible for the ageing of local places in other ways. Ageing is, in part, a function of technological advances that have increased human longevity through improvements in health care and support. Such technological advances are an outcome of an economic system that frees people from routine, manufacturing jobs; this encourages research and development programmes, which assist industrial development, and often have spin-offs for human services. Technological change has other effects as well. The 'decline' of the family farm is one outcome of the mechanization of agriculture that has reduced year-round labour demands. Cheap, seasonal labour can move in and out of agricultural regions as demand warrants. That is, there are fewer permanent employment opportunities for the younger generation of rural residents. Economic restructuring has, therefore, been responsible for changes that have encouraged the ageing of rural areas. Thus, an exploration of the geography of ageing reveals that it is a product of economic restructuring, not just a cause.

Conclusions: A Geography of Rural Ageing?

The elderly are an important part of the social structure of rural regions, both in terms of their absolute and relative numbers. The concentration of this older population is the outcome of both demographic and economic change. Some old people choose to retire to rural communities that offer attractive amenities and thus in-migration increases the local elderly population. In other places, however, the out-migration of young people has contributed to the local growth of senior citizens. Whether or not this group is a 'burden' on local resources really depends on their financial status. The relatively affluent in-migrants bring with them resources that may not be available to the less healthy and wealthy individuals who do not move in old age.

After reviewing the literature, it must be admitted that, given the importance of the ageing population, the degree of geographic attention to the question of the rural elderly in the US and UK is not substantial. Topics relating to rural issues and the elderly population are marginalized within the discipline. The rural elderly thus face a process of double marginalization. In part, this results from the limited attempts at understanding the broader context in which rural ageing occurs.

Rural ageing must be seen as intimately linked to the broader process of social and economic restructuring that are discussed in the other chapters of this volume. For example, the links between rural amenities, tourism and retirement are worthy of closer scrutiny; the displacement of farm labour has driven young people away from rural places and thus has promoted the ageing of these communities; the plight of rural workers who survive their husbands raises the prospects of extending Mackenzie's analysis of farm women (Chapter 5); and, of course, any discussion of the provision of rural housing (see Robinson, Chapter 7) must include a consideration of the housing needs of an ageing population.

Given the broad base of research provided by rural social gerontologists, there is a wide arena of research ready for geographers to delve into and expand upon.

References

Aitian, P. and Aitian, R. (1990) Health and social antecedents of relocation in rural elderly persons. *Journal of Gerontology* 45, 532–38.

Bealer, R., Willits, F.K. and Kuvelsky, W.P. (1965) The meaning of rurality in American society: some implications of alternative definitions. *Rural Sociology* 30, 255–56.

Bender, L., Green, B.L., Hady, T.F., Kuehn, J., Nelson, M., Perkinson, L. and Ross, P. (1985) The diverse social and economic structure of non-metropolitan America. *Rural Development Research Report* 49, Department of Agriculture, Economics, Statistics and Co-operative Farmers, Washington DC.

Bohland, J. and Rowles, G. (1988) The significance of elderly migration to changes in the elderly population concentration in the United States: 1960–1980. *Journal of Gerontology* 43, 145–52.

Bryant, E.S. and El-Attar, M. (1984) Management and redistribution of the elderly – a challenge to community services. *Gerontologist* 24, 634–40.

Cebula, R. (1974) The quality of life and migration of the elderly. *Review of Regional Studies* 4, 62–68.

Clifford, W., Heaton, T. and Fuguitt, G. (1985) Residential mobility and living arrangements among the elderly: changing patterns in metropolitan and non-metropolitan areas'. *International Journal of Aging and Human Development* 14, 139–56.

Coward, R.T. and Lee, G. (eds) (1985) *The Rural Elderly in Rural Society: Every Fourth Elder*. Springer, New York.

Cribier, F. (1980) A European assessment of aged migration. *Research on Aging* 2, 255–70.

Cross, D. (1990) *Counterurbanisation in England and Wales*. Gower, Aldershot.

Crown, W. (1988) State economic implications of elderly interstate migration. *The Gerontologist* 28, 533–39.

Dunbar, G. (1990) Retiree-based economic development. *The Business Report* (July), Pennsylvania Chamber of Business and Industry, Hamburg, PA, pp. 4–5.

Fuguitt, G. and Tordello, S. (1980) Elderly net migration: the new trend of nonmetropolitan population change. *Research on Aging* 2, 191–204.

Gant, R. and Smith, J. (1988) Journey patterns of the elderly and disabled in the Cotswolds: a spatial analysis. *Social Science and Medicine* 27, 173–78.

Gant, R. and Smith, J. (1991) The elderly and disabled in rural areas: travel patterns in the North Cotswolds. In: Champion, T. and Watkins, C. (eds) *People in the Countryside*. Paul Chapman Publishing, London, pp. 108–124.

Gober, P. and Zonn, L. (1983) Kin and elderly amenity migration. *The Gerontologist* 23, 288–94.

Golant, S. (1987) Residential moves by the elderly persons to US central cities, suburbs-rural areas. *Journal of Gerontology* 42, 534–39.

Golant, S. (1990) Post 1980 regional migration patterns of the US elderly population. *Journal of Gerontology* 48, 135–40.

Golant, S., Rowles, G. and Meyer, J. (1988) Aging and the aged. In: Gaile, G. and Willmott, C. (eds) *Geography in America*. Merrill Publishing Co., Columbus, pp. 451–66.

Graff, T. and Wiseman, R. (1990) Changing pattern of retirement counties since 1965. *Geographical Review* 80, 239–51.

Grundy, E. (1987) Retirement migration and its consequences in England and Wales. *Ageing and Society* 7, 57–82.

Harper, S. (1987) The kinship network of the rural aged: a comparison of the indigenous elderly and the retired in-migrant. *Ageing and Society* 7, 303–27.

Hiltner, J., Smith, B. and Sullivan, J. (1986) The utilization of social and recreational services by the elderly: a case study of Northwestern Ohio. *Economic Geography* 62, 232–40.

Hugo, G. (1984) The ageing of ethnic populations in Australia. NRIGGM, University of Melbourne, Melbourne.

Jones, D., Victor, C. and Vetter, N. (1985) The problem of loneliness in the elderly in the community. *Journal of the Royal College of General Practitioners* 35, 136–39.

Karn, V. (1977) *Retiring to the Seaside*. RKP, London.

Kivett, V. (1988) Aging in a rural place; the elusive South of Well-Being. *Journal of Rural Studies* 4, 125–32.

Krout, J. (1983) *The Rural Elderly*. Greenwood Press, Westport.

Krout, J. (1984) *The Utilization of Formal and Informal Support of the Aged: Rural Versus Urban Differences*. Fredonia; Final report to the American Association of Retired Persons, Andrus Foundation, New York.

Krout, J. (1986) *The Aged in Rural America*. Greenwood Press, Westport.

Krout, J. (1988) The elderly in rural environments. *Journal of Rural Studies* 4, 103–14.

Lassey, M. and Lassey, M. (1985) The physical health status of the rural elderly. In: Coward, R.T. and Lee, G. (eds) *The Rural Elderly in Rural Society: Every Fourth Elder*. Springer, New York, pp. 83–104.

Law, C. and Warnes, A. (1976) The changing geography of the elderly in England and Wales. *Transactions of Institute of British Geographers* NS1, 453–71.

Longino, C. (1990) Geographical mobility and family caregiving in nonmetropolitan America: three decade evidence from the US Census. *Family Relations* 39, 38–43.

Longino, C., Wiseman, R., Biggar, J. and Flynn, C. (1984) Aged metropolitan-nonmetropolitan migration streams over three census decades. *Journal of Gerontology* 39, 721–29.

McLaughlin, D. and Jensen, L. (1991) Poverty among the elderly: a metro-nonmetro comparison. Paper presented at the Rural Sociological Society Annual Meeting, Columbus, Ohio.

Meyer, J. (1981) Equitable nutrition services for the elderly in Connecticut. *Geographical Review* 71, 311–23.

Meyer, J. (1987a) County characteristics of elderly net migration rates: a three-decade regional analysis. *Research on Aging* 9, 441–52.

Meyer, J. (1987b) A regional scale temporal analysis of the net migration patterns of elderly persons over time. *Journal of Geography* 42, 366–75.

Meyer, J. and Kern, R. (1990) Economic and social implications of demographic change. In: Fossler, R.S., Alonso, W., Meyer, J.A. and Kern, R. (eds) *Demographic Change in the American Future*. University of Pittsburgh Press, Pittsburgh, pp. 79–132.

Meyer, J. and Speare, A. (1985) Distinctively elderly migration: types and determinants. *Economic Geography* 61, 79–88.

Mosely, M. (1978) The mobility and accessibility problems of the rural elderly. In: Garden, J. (ed.) *Solving the Transport Problems of the Elderly*, Beth Johnson Foundation, University of Keele, pp. 51–62.

Nelson, G. (1980) Social services to the urban and rural aged: the experience of area agencies on aging. *The Gerontologist* 27, 200–7.

Newhouse, J. and McAuley, W. (1987) Use of informal in-home care by rural elders. *Family Relations* 36, 456–60.

OECD (1988) *Ageing Populations: the Social Policy Implications*. OECD, Paris.

Ormond, R. (1986) Recent interstate net migration flows of the elderly in California. *The California Geographer* 26, 45–57.

Rogers, A. and Watkins, J. (1987) General versus elderly interstate migration and population redistribution in the United States. *Research on Aging* 9, 463–529.

Rowles, G. (1981) The surveillance zone as meaningful space for the aged. *The Gerontologist* 21, 304–11.

Rowles, G. (1983a) Between worlds: a relocation dilemma for the Appalachian elderly. *International Journal of Aging and Human Development* 1, 301–14.

Rowles, G. (1983b) Place and personal identity in old age: observations from Appalachia. *Journal of Environmental Psychology* 3, 299–313.

Rowles, G. (1986) The geography of aging or the aged: toward an integrated perspective. *Progress in Human Geography* 10, 511–39.

Rowles, G. (1988) What's rural about rural aging? An Appalachian perspective. *Journal of Rural Studies* 4, 115–24.

Serow, W. (1987) Determinants of interstate migration: differences between elderly and nonelderly movers. *Journal of Geography* 49, 95–100.

Sofranko, A., Williams, J.D. and Fliegel, F. (1981) Urban migrants to the Mid-

West: some understandings and misunderstandings. In: Roseman, E.E. *et al.* (eds) *Population Redistribution in the Midwest.* Iowa State University Press, Ames, pp. 97–127.

Sofranko, A., Fliegel, F. and Glasgow, N. (1983) Older urban migrants in rural settings: problems and prospects. *International Journal of Aging and Human Development* 16, 297–309.

Speare, A. and Meyer, J. (1988) Types of elderly residential mobility and their determinants. *Journal of Gerontology* 43, 574–81.

Steinhauer, M. (1980) Obstacles to the mobilization and the provision of services to the rural elderly. *Educational Gerontology* 5, 399–407.

US Bureau of the Census (1989) *Statistical Abstracts of the United States, 1989.* 109th edition, US Bureau of the Census, Washington DC.

Victor, C., Vetter, N. and Jones, D. (1983) The contribution of sheltered housing accommodation to the housing of the elderly in a rural and an urban area of Wales. *Journal of Housing for the Elderly* 1, 19–28.

Warnes, A. and Law, C. (1984) The elderly population of Great Britain: locational trends and policy implications. *Transactions of Institute British Geographers* NS9, 37–59.

Warnes, A. and Law, C. (1985) Elderly population distributions and housing prospects in Britain. *Town Planning Review* 56, 282–313.

Watkins, J. (1990) Appalachian elderly migration: patterns and implications. *Research on Aging* 12, 409–29.

Wenger, C. (1984) *The Supportive Network: Coping with Old Age.* George Allen and Unwin, London.

Wenger, C. (1991) Survivors: support network variation and sources of help in rural communities, *Journal of Cross-Cultural Gerontology* 6, 41–82.

Williams, J. and Sofranko, A. (1979) Motivations for the immigration component of population turnaround in nonmetropolitan areas. *Demography* 16, 239–55.

Wiseman, R. and Roseman, C. (1979) A typology of elderly migration based on the decision making process. *Economic Geography* 55, 324–37.

7

THE PROVISION OF RURAL HOUSING: POLICIES IN THE UNITED KINGDOM

Guy Robinson

In the last decade, three issues have attracted particular academic and public attention with respect to the provision of housing in rural areas of the United Kingdom (UK). These are a lack of housing for those on low incomes, and especially a shortage of rented accommodation, the changing character of government support for housing, which has helped reduce investment (Shucksmith, 1984; MacGregor *et al.*, 1987), and the extent to which new housing should be located on 'greenfield' sites. Together these issues have tended to replace the earlier more diverse focus upon problems associated with the presence of concentrations of inadequate housing in peripheral areas such as northeast Scotland, the Scottish islands, mid-Wales and Cornwall. In some respects this shifting focus reflects improvements in the quality of rural housing which has, therefore, diverted attention towards issues of housing provision. For example, in the 1940s 11.6% of rural houses in the UK were found to be 'unfit for occupation' and 33.4% 'in need of repair' (Ministry of Housing, 1944, 1948), whereas three decades later the proportion deemed unfit had fallen to around 5% (Gilg, 1985, pp. 60–1; Rogers, 1983). However, changes in the nature of research carried out on rural housing have also shifted the focus towards issues associated with social justice and the political background to housing provision and management (Phillips and Williams, 1983; Shucksmith, 1990, p. 18).

At a time when central government policies can be seen to be having a significant impact on the delivery and nature of rural housing provision, there has been a gradual response from rural researchers who have followed their urban counterparts in pursuing work founded in both Weberian and Marxist traditions (for example, Bassett and Short, 1980, p. 2; Cloke, 1989). This work, with its stress upon constraints within the housing market, the nature of housing allocation and the reproduction of the labour force, has begun to reveal the particular characteristics of housing problems in rural areas and to build upon important work done on the role of landownership and large estates in rural economy and society (for example, Massey, 1977; Massey and Catalano, 1977; Newby *et al.*, 1978; Munton and Goodchild, 1985).

This chapter investigates the changing context of rural housing problems in the UK, focusing on key issues of the growing problem of housing shortages in rural areas, links between these shortages and changes in housing policy, and the debate over the need for substantial development outside existing settlements. It is suggested that central government may attempt to resolve some of these housing problems by effecting changes in planning policy, and that this is likely to provide a significant context for research during the 1990s. Some suggestions are given as to the possible focus of this research.

No Homes for Locals

As prices inexorably rise, so the population which actually achieves its goal of a house in the country becomes more socially selective. Planning controls on rural housing have therefore become – in effect, if not in intent – instruments of social exclusivity.

(Newby, 1980, p. 187)

This exclusivity, in which more people are seeking a rural idyll and purchasing homes in the countryside, whilst others are denied access to housing in their local rural area, is one factor in the escalation of a growing problem of housing shortages in rural areas. Whereas an increasing number of urban residents are moving to homes in the countryside, and maybe even a second home, long-term rural residents and their offspring are experiencing increasing difficulty in finding suitable rural accommodation, either to purchase or rent (for example, Sherwood, 1991). This is the 'no homes for locals' issue, which is one of the most alarming and emotive problems of a number of difficulties that can be considered under the umbrella of rural housing problems.

Several arguments have been proposed for the introduction of policies favouring the allocation of housing for local inhabitants (Table 7.1). These have often proved sufficiently powerful to overcome any counter-arguments suggesting that the need to cater for locals only is a poor foundation for policy. If taken to extremes, with strong enforcement of exclusion of outsiders, as in the Channel Isles, then such policies could lead to stagnation and social atrophy in rural communities. Nevertheless, the plight of certain groups in rural society being denied access to housing in their 'home area' is an emotive one, underlain by a real need to which policy-makers have had to respond.

Special attention has been given to the issue in those areas where

Table 7.1. Arguments in favour of policies allocating housing for local inhabitants.

1. Locals are frequently equated with the poor and deprived in rural society, though this ignores the presence of wealthy elements
2. The possibility that outsiders might have grounds for housing in a particular rural locality are overlooked because of 'prior claims' of locals
3. Housing locals is seen as helping to maintain employment and services, though the fact that housing outsiders too might have the same impact is often ignored
4. Elected representatives such as parish and county councillors feel that they are morally obliged to press the claims of locals to be housed locally
5. Championing the plight of 'homeless locals' has a media appeal that puts rural housing issues onto the public and political agenda

Source: based on Rogers (1985a, pp. 375–76).

outsiders have turned a significant proportion of the housing stock into second homes, although this is not so widespread a phenomenon as in many European countries. Second home owners are often in direct competition for property with local first-time buyers or renters. So locals can be out-bid for property, or prices can be increased by this particular influence on the market, again to the detriment of the locals. In some cases this has provoked conflict between the newcomers and local inhabitants, some of the most extreme examples occurring in Wales where it has been exacerbated by the cultural factor of the perceived threat felt by locals to their Welsh language and culture (Bollom, 1978; Grant, 1991). Similar conflicts have occurred on Skye between Gaelic-speaking locals and English second home owners (Whyte, 1978). Another problem has been the greater cost of provision of utilities in areas of second home ownership, these costs being partly transferred to permanent residents possibly not so affluent as their new neighbours. However, benefits associated with the presence of second homes have also been apparent: notably the retention of services and stimulus to the local economy, especially in areas experiencing long and continued depopulation (Clout, 1969, 1971; Pacione, 1979; Shucksmith, 1983).

In the late 1970s, problems over the high incidence of second home ownership in the Lake District were sufficiently great for a proposed ban to be contemplated upon further second home buyers (Clark, 1982a; Shucksmith, 1981). Between 1977 and 1984 the Lake District Special Planning Board (LDSPB) operated a policy to exclude outsiders from purchasing new housing stock in the area, though this exclusion did not apply to purchasers of existing housing: 'The Board will seek means by which all further housing development can be retained for occupation by local people as full-time residents, except in the case of divisions of existing dwellings or where exceptional site factors apply' (LDSPB, 1980, p. 55). This meant that the influx of outsiders in general and second home owners in particular was not effectively controlled by the policy (Capstick, 1987;

Shucksmith, 1990, pp. 107–37) and this weakness contributed to the Secretary of State for the Environment ordering the Board to delete the policy from its structure plan. However, a similar policy has recently reappeared in the Peak District National Park, though here it is not just aimed at keeping out second home owners but at restricting any forms of new development within the park, with the exception of new low-cost houses for locals.

The problem of the lack of accommodation for the rural working class and lower middle class is not a new one. Rural areas are typified by a higher incidence of owner-occupier housing than urban areas and, because of this, housing supply is more controlled by the market. This can create problems for low-income groups, especially if they are competing for housing in communities where competition is heightened by demand for second homes, and homes for retirement migrants, long-distance commuters and the emergent 'service class' (Thrift, 1987). Traditionally, low-income groups have tended to rely on the public and private rented sectors. However, the latter has declined markedly as tied cottages, formerly occupied by farmworkers, have been taken out of this sector (Lewis and Bowler, 1987). The decline in farm labour has prompted the sale of these cottages, either to the sitting tenants or on the open market or even as holiday accommodation. This has placed more pressure on the public rented sector, exposing shortcomings that can often be attributed to the lack of significance attached to such provision by local authorities. Traditionally, landowners preferred to house their rural labour force in tied accommodation as a means of reinforcing worker dependency, and Newby (1980, pp. 184–86) has argued that this attitude towards housing provision has influenced rural local authorities where political control has often rested largely with farmers and landowners. Thus the political will to embark on major local authority (council) house construction programmes has not existed and, moreover, the landowners, as major ratepayers, have been reluctant to incur the extra costs such programmes might bring. Following the *Housing Act* of 1980, and the so-called 'right to buy', sales of council houses have further exacerbated the problem, in many cases rates of such sales being higher in rural areas than urban areas (Foulis, 1985).

Sales of Local Authority (Council) Housing

The public rented sector controls a substantial element of the UK housing stock – just under one-third of all households. But from the mid-1970s, a number of measures have been taken by central government to limit the sector's extent. In 1977 the Labour Government introduced controls on public expenditure through Housing Investment Programmes (HIP) which set a maximum amount of money to be spent on capital works by each local authority. This shift towards centralization of decision-making, on the appropriate level of government investment in housing, reflected concerns over the general quality of council housing and its management, as well as cost considerations (Gibb and MacLennan, 1986). Controls were then intensified from 1979 under the Conservatives as part of their policy of 'balancing the budget' and constraining monetary growth. Municipal intervention was discouraged in favour of deregulation, privatization and competition. There were major reductions of HIP allocations and sales of council houses at a discount to tenants. Furthermore, councils' reinvestment of funds from sales was restricted: to 50% of such funds being available for new house-building and then, from 1985, to just 20%. Central government regulation of local authority housing was extended by controlling authorities' capital borrowing, by limiting the housing support grant and, recently, by controlling the rate of Poll Tax levied. The 'de-municipalization' of the housing stock was carried further in the 1988 *Housing Act*, in which council tenants were given the right to opt for a new landlord within the private sector (Malpass and Murie, 1987, p. 101).

To some extent the potential problem of high levels of sales of council houses in 'desirable' rural locations was recognized by some limited restrictions in the 1980 *Housing Act*. The Act disqualified people who had not lived in the area for more than 3 years from purchasing former council houses in certain rural areas. There have also been some attempts by local planning authorities to operate policies favouring 'local need'. In national parks, Areas of Outstanding Natural Beauty and areas granted special status, a covenant can be attached to council houses to restrict sales to locals only, or to give the local authority first refusal on resale over a 10-year period. However, of 130 authorities applying for special status, only 18 have been successful (Shucksmith, 1990, p. 54). Fewer restrictions on sales have been permitted in Scotland, where they can only be enforced once one-third of the stock has already been sold (Shucksmith, 1988b).

It has been argued that unrestricted council house sales may have long-term detrimental consequences, in that they restrict the chances of rehousing families locally and so can threaten the viability of small communities (Phillips and Williams, 1981; Williams and Sewel, 1987, p. 72). Two areas in particular may have high rates of sales and may be associated with special types of problems:

1. Areas, such as the Lake District, which have great scenic beauty, are popular with tourists and second home owners, and where there are high demands for residential property.
2. Areas within commuting distance of major urban areas with expanding labour markets, again with high pressure upon housing.

In both cases a process of rapid 'residentialization' of public sector housing is occurring, so that 'both the quality and size of the stock is reduced and the tenantry include an increasing proportion of marginalized groups such as the elderly, the unemployed, single parents, and the sick and disabled' (Williams and Sewel, 1987, p. 73). This reflects the fact that local authorities are selling their best stock to middle-aged tenants possessing some capital, with the result that the stock is depleted both in quantity and quality.

The changing policies towards local authority housing contributed to a 40% increase in waiting lists for rural council houses in Scotland from 1981 to 1986 (Fig. 7.1a), an 83% rise in homelessness from 1983 to 1987 in rural Scotland (Fig. 7.1b), and higher levels of houses below tolerable standards. These problems have been compounded by problems of access to suitable homes for first-time buyers, the elderly and those requiring sheltered housing (Mooney, 1991). Furthermore, sales of council houses have reduced the availability of public sector housing, whilst private sector property for renting has also declined. Investment in council housing in real terms fell by two-thirds between 1976 and 1986 (see Shucksmith, 1988a) (Fig. 7.2), and there continued to be a strong urban bias in housing policy.

It is important, though, to view the sales of local authority houses in the context of other changes in government housing and planning policy during the 1980s. In particular, moves towards establishing new sources of accommodation for rent and new low-cost houses for sale need closer scrutiny, as do changing attitudes to planning constraints that affect the rural housing market.

Policy Solutions in the 1980s

By scaling down the amount of access to local authority housing at a time when the private rented sector has also been contracting, the government has shifted the burden of catering for housing needs to private developers and organizations such as housing associations.

A consultative document produced by the Department of the Environment (DoE) (1977) in the late 1970s envisaged that housing associations would become the 'third arm' of housing provision, alongside owner-occupation and local authority housing, and thereby replace the declining private rented sector. Therefore, the role of housing associations is seen as crucial in the provision of 'social housing' in which the financing, provision and management of property is carried out by a number of different organizations, including special associations, cooperatives, building societies and trusts. The Conservative Government has favoured greater involvement of private financing of this social housing (see Clark, 1988), though most housing associations are funded mainly by central

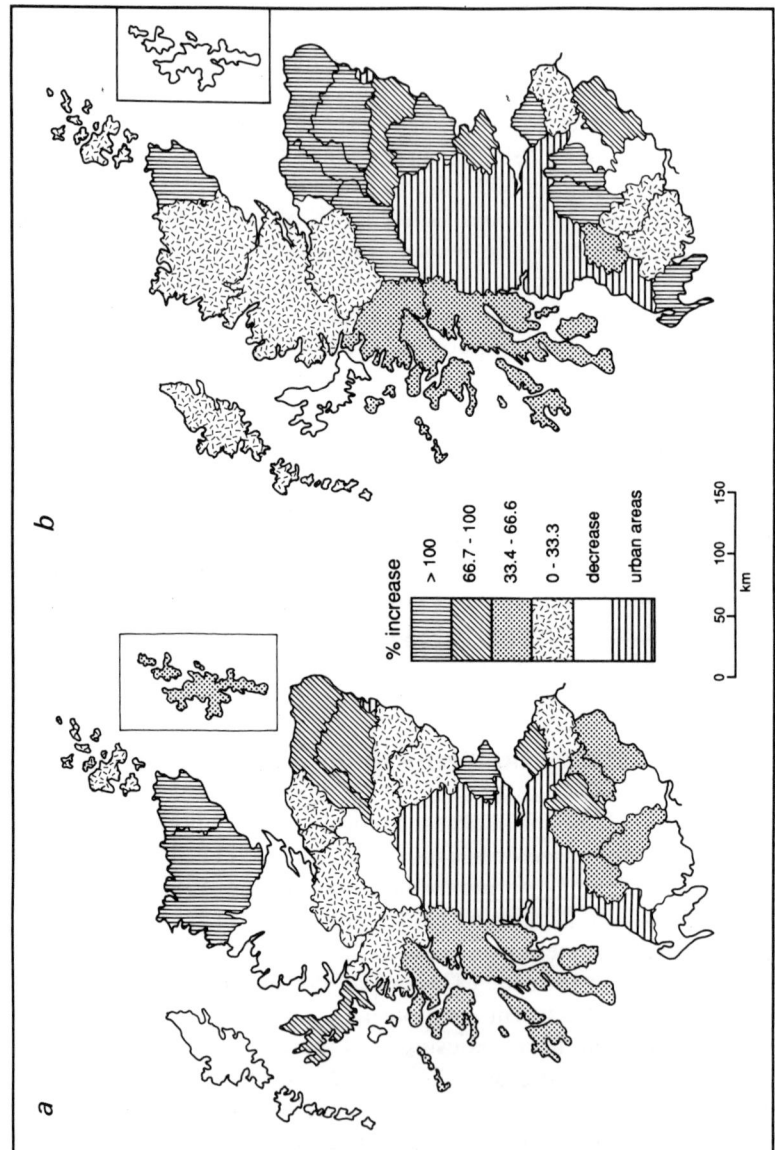

Fig. 7.1. (a) Percentage change in waiting lists for council houses in rural Scotland, 1981–1986; (b) Percentage change in homelessness in rural Scotland, 1983–87 (Source: after Alexander *et al.*, 1988).

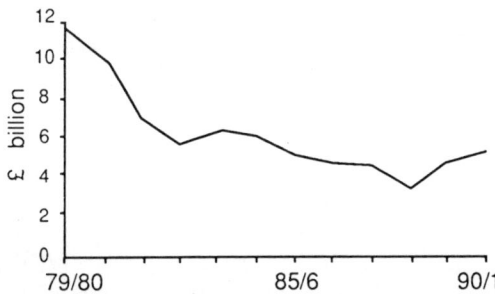

Fig. 7.2. Government spending on housing in real terms in Great Britain, 1979–1991 (Data source: Shelter).

government through the Housing Corporation (HC), the quango that funnels public money to housing associations. During the 1980s central government has increasingly promoted the development of housing associations, for example favouring them and village housing trusts to provide low-cost accommodation in the countryside (Conservative Policy Centre, 1989). This solution to the problem of rural housing need envisages the promotion of trusts and associations nationwide to stimulate small-scale development that will meet local needs as determined by local surveys. This approach has been supported by the Rural Development Commission (RDC) and the DoE who, in 1988, doubled the funding to village housing schemes by the HC. A prime aim of this approach is that rural low-income families should have first claim to new housing which would be available at low rents. Furthermore, with schemes for easy purchase too, local landowners could be offered a stake in the resale value of homes built by trusts, thereby encouraging them to provide land. More encouragement for such involvement is provided in the Business Expansion Scheme operating in rural areas, which provides investors in housing with generous relief from income and capital gains tax for a 5-year period. Furthermore, the 1988 *Housing Act* specified that all new tenancies were to be at open market rents for assured tenancies, therefore aiming to decontrol rents.

It is not surprising, therefore, that housing associations are increasingly playing the prime role in new public sector rented housing: contributing 19 000 new houses in 1990, compared with 13 000 new council houses out of a total of 184 000 new houses (the lowest total since 1982) (Fig. 7.3). However, to date, the work of housing associations in rural areas has been limited in comparison with that in urban areas. In Scotland, for example, their work has focused on special-purpose housing, such as sheltered accommodation, and their activities have been directed mainly at urban housing problems (Shucksmith and Watkins, 1989).

Government support for housing associations in rural areas echoes sentiments expressed in previous policies but which have either not been translated into workable practicalities or have been implemented more

Guy Robinson

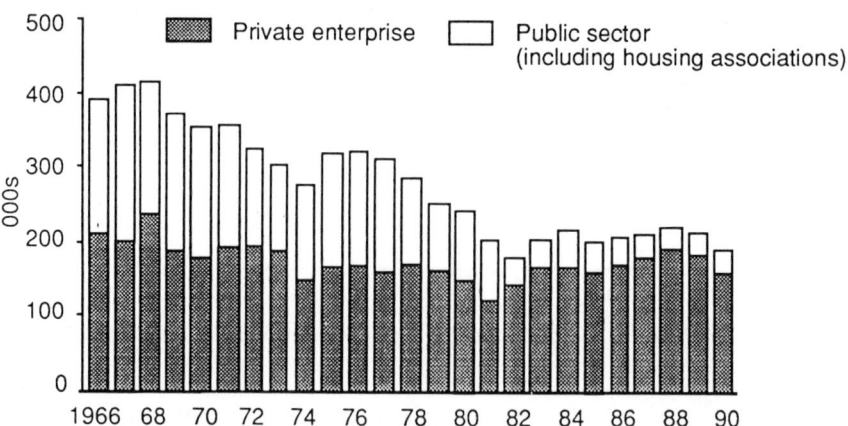

Fig. 7.3. Housing completions in Great Britain, 1966–1990 (Data source: Housing and Construction Statistics).

slowly in rural areas. For example, only 3% of the budget of the HC is devoted to small-scale housing schemes in the countryside. It seems clear, therefore, that a more determined nationwide policy encouraging low-cost rural housing is required, perhaps building on some of the local initiatives of community trusts, but without the limiting constraints currently imposed by central government that affect rentals and the degree of control that housing associations can exert. In Scotland a significant development may be the establishment of Scottish Homes, a separate Scottish arm to the HC, which is specifically charged with extending housing association activities into rural areas.

The growing role of private organizations in attempting to provide affordable housing for rural residents on modest incomes is illustrated by the National Agricultural Centre Rural Trust (NACRT), established in 1975. This is a charity that has the support of the Royal Agricultural Society of England and the Rank Foundation. It pursues its objectives by helping housing associations to build in villages, and its particular role is to advise and assist village groups, such as parish councils, to identify housing needs and sites for development, working in conjunction with the RDC and the House Builders Federation (Rogers, 1985a). Under the slogan 'Village Houses for Village People', the NACRT has also established 14 charitable associations which aim to build small groups of houses, flats and cottages (NACRT, 1987). These buildings will not be sold, but will be let to particular groups within rural society, notably young adults wishing to remain in the countryside, and the elderly who wish to move from larger accommodation but still remain in their local rural area.

The NACRT requires cheap land in order to pursue its aims. Given the general lack of such land in areas designated for new housing, it often purchases land adjacent to existing settlements but not scheduled for development. It then lobbies the local authority for zoning to be altered, with the proviso that the new housing will go to locals or to specially identified groups.

The NACRT has been supported by the RDC who recognize that housing initiatives must be integral to its development strategy for rural areas. The RDC has encouraged the HC to make special allocations for rural areas and has funded field workers working for the NACRT, though this is on a very small scale (Rogers, 1985a, 1987). In its earlier form, as the Development Commission, it cooperated with the HC to build houses in areas where advance factories were being constructed and, in a later scheme, to build houses in 16 villages on a shared ownership basis. However, both of these projects revealed weaknesses that have generally tended to affect the operations of rurally based housing associations. These problems have included difficulties in attracting funds away from urban areas or 'urgent housing needs', high construction costs per house because of limited economies of scale in small developments, difficulties identifying 'needy' local groups, restricting costs so that the new houses remain in the price range of first-time buyers, and preventing an influx of outsiders through instituting restrictive clauses on resales (Shucksmith, 1990, pp. 160–1).

Recently, further support from the HC for the NACRT has come for new rural houses to be made available for rent in developments as small as four or five houses, and with new county-based housing associations established to administer such schemes (NACRT, 1987, pp. 37–8). Of houses built under this arrangement, 25% have been for the elderly and disabled, and the remainder for young couples and families. In order to keep rents low, 90% of capital costs are being provided by the HC in the form of a Housing Association Grant.

Some attempts have been made to provide workshops and houses in the same development, for instance, the RDC's Craft Homes Experiment (EOTECH, 1989). Unlike the RDC's emphasis on housing in 'remoter' rural areas, though, the great majority of these comparable schemes are in towns and villages in well-populated countryside. Certainly, the relative success of some of these schemes does suggest a potential avenue for future rural housing projects to consider: namely, the integration of housing and employment needs in planning that extends beyond narrow sectoral concerns.

Support for this, and other changes, in planning is implicit in some of the present government's general philosophies on the importance of 'free market' forces and the removal of restrictive regulations. In this respect, one possibility would be a reduction in planning controls, as proposed in the 1985 White Paper, 'Shifting the Burden', and as manifest in the Department of Trade and Industry's support for new village schemes at Foxley Wood (Hampshire) and Stone Bassett (Oxfordshire), amongst a number of 'new village' proposals (Bell and Cloke, 1989; Shucksmith, 1990, pp. 8–10; Wilcox, 1991). The sharp rise in house prices in the 1980s, especially in the southeast, has also put additional pressure on planning controls in other regions, with demand for housing being pushed to areas where cheaper land could be made available for housebuilding. This raises the question of the extent to which new housing development outside existing settlements (greenfield sites) will be restricted in future.

'Greenfield' Development

One manifestation of government support for reduced planning controls is in the encouragement of local councils to set aside land specifically for new low-cost homes for rural residents. Although outside the area designated for development in structure plans, this land can then become part of village infill, but so designated that developers could be prevented from building high-cost housing on it. This is just one aspect of a number of issues affecting the perceived need to build outside existing settlements, with conflicting arguments presented as to the need for such development.

The problems associated with the increased demand to build new houses in the countryside – to accommodate commuters and former urban residents settling in 'deeper' countryside – have provoked an increasingly acrimonious debate (Clark, 1982b). The conflict has been at its sharpest in southeast England, where those advocating more new housing developments have argued for reduced planning controls on such schemes, especially within the Metropolitan Green Belt. Different attitudes to controls have partly reflected variations in the estimates of how many new houses are required (Robinson, 1990, pp. 390–1). Higher estimates have increased concerns that further relaxations in Green Belt controls will be encouraged in order to accommodate the large scale of development predicted. Alternatively, there is a fear that the Green Belt will be 'leap-frogged', with new housing schemes pushed towards the outer fringes of the southeast region. This is already happening to the west of London in the so-called 'M4-corridor', in Berkshire, and hence the growing claims that the south-east is being engulfed by a tide of concrete, bricks and mortar. In particular, spurred on by the prospect of economic growth to be promoted by the transport corridor from London through Kent to the Channel Tunnel, major development schemes are likely to be proposed to the south and south-east of London (see Blunden and Curry, 1988, pp. 92–3).

The demand for more housing in close proximity to major urban areas has placed growing pressures on land within Green Belts. In the past such plans for large-scale greenfield development have conflicted with long-held views that preservation of agricultural land must remain paramount.

Hence an important change in the general planning environment, which may affect rural housing provision, is the erosion of this longstanding view that as much agricultural land as possible must be retained and protected from any building development. This changing attitude is illustrated in the Alternative Land Use and the Rural Economy (ALURE) package introduced in 1987. This contained proposals from the Ministry of Agriculture, Fisheries and Food (MAFF) to encourage farm diversification through alternative uses of farmland, and especially the expansion of 'environmentally friendly' farming. The package incorporated additional proposals from the DoE which some felt would encourage building development on good farmland by reducing MAFF's input to planning controls (Blowers, 1987; DoE, 1987). The DoE's proposals are that promotion of economic activity should be considered, as well as protection of the countryside, when evaluating applications for planning permission affecting agricultural land. This could lead to a reversal of previous presumptions against development on land in agricultural use (Robinson, 1991, p. 97).

If more agricultural land is released for house building, then existing rural house-owners are likely to be among those protesting about such 'destruction' of the countryside, partly because this development can counter tendencies towards increases in property value that accrue once restraint on development is in place (Herington, 1984, p. 152; Hoggart and Buller, 1987, pp. 223–5). In effect, owner-occupiers can gain by restrictions on the supply of new house construction as restrictions inflate the exchange value of their own houses (Pratt, 1982). Hence, Herington (1984, p. 152) asserts that, 'locally-based pressure groups usually emerge when specific development proposals threaten the property values of local residents.' In contrast, rural dwellers who do not own domestic property are unlikely to gain from restricted development. Conversely, they may gain from increased house-building, especially if this provides opportunities to move from rented property to owner-occupier housing, and/or if it reduces rents and house prices (Shucksmith, 1987). Therefore, there is a strong implication that the operation of rural property markets promotes class conflicts, with a cleavage based on the nature of housing tenure. The greater political power wielded by owner-occupiers within the local state has helped promote restraint upon development, especially via planning controls exercised by county and district planning authorities (Short et al., 1987, p. 41). However, increasingly, central government has diminished local planning authorities' control on development by using the appeals procedure to overrule local decisions. So the local planning authorities face the dilemma of either designating more land for development or having their constraints on development overruled on appeal. This practice of developers utilizing the appeals procedure has grown significantly during the 1980s. Their success rate has also risen: from 25% in 1977 to 41% in 1987.

One response to this apparent loss of control over development that the local authorities have experienced has been 'private' bargaining between the authorities and developers. This permits development, but controls its form by requiring developers to provide or pay for certain amenities and services that may not have been included in the original planning application (Short *et al.*, 1987, p. 49; Reade, 1987, p. 91). Therefore, new residential development is not prevented but its form may be altered, or maybe the developers' preferred location is 'sacrificed' in return for an alternative location.

In reaching agreements with developers, it is often the case, though, that local authorities pay insufficient attention to the visual effects the new houses may have upon small rural communities. For example, Aslet (1991) argues that many new developments have irrevocably altered the appearance of villages. This alteration has involved the construction of too many new houses out of character with previous development, frequently of dubious architectural value and concentrated in particular villages 'designated for growth', rather than being spread more thinly among a number of communities. So there has been established an unsatisfactory mix-and-match of house styles for those wealthy enough to purchase new 'executive' homes in the countryside. Meanwhile, examples of sensitive planning involving the use of distinctive local styles have become rare. Furthermore, the occupants of the new houses have often been former urban residents who have not required local services, so that an increased village population has not meant a reprieve for rural schools, buses, shops and other services.

Conclusion

For researchers on rural housing problems, the issues referred to above offer great scope for more concerted and systematic investigations than have occurred previously. This scope extends well beyond the obvious class conflicts in the 'no homes for locals' and greenfield development problems, and also beyond the analysis of changes in government housing policy. There is a need to investigate the relationships between general economic changes, institutions, external policy constraints and the housing market in rural areas. Such work must recognize that the overall structure of political and economic power cannot be 'taken for granted', and that agencies such as local housing authorities and housing associations operate in what McDowell (1982, p. 82) refers to as 'an exogenously determined environment'. It is the nature of the impacts of this environment upon the housing market that provides the essential setting for any proper development of an understanding of rural housing problems. Therefore, although the inherent conflict between the needs of housing provision and the desire to preserve the landscape will continue to provide a strong research focus, this focus will include explicit recognition of the role of institutions and external policy constraints.

Given that many of the key policies affecting housing have been part of land-use planning, it is changes in this planning that are likely to form a starting point for much research. The current attitude that favours fewer government controls is supported by trends towards the maintenance of less agricultural land and calls for relaxation of planning controls to permit more house-building. The complex interaction between any such changes in land-use planning, land prices, taxation policy, mortgage finance and the agencies involved in building new housing will continue to provide scope for work founded in a number of research traditions.

The growing awareness of the extension into rural areas of problems of homelessness and inadequate housing stock should also provide scope for both research and new solutions to these problems. It is to be hoped that policy-makers and academics alike will extend this scope to encompass examples from other countries. 'Self-help' schemes, and an even greater role for non-governmental finance, may play a vital part in dealing with future housing problems in the UK, depending on the central political direction established.

References

Alexander, D., Shucksmith, M. and Lindsay, N. (1988) *Scotland's Rural Housing: Opportunities for Action*. Shelter (Scotland)/Rural Forum, Edinburgh and Perth.

Aslet, C. (1991) *Countryblast*. John Murray, London.

Bassett, K. and Short, J.R. (1980) *Housing and Residential Structure: Alternative Approaches*. Routledge & Kegan Paul, London.

Bell, M. and Cloke, P.J. (1989) The changing relationship between the private and public sectors: privatisation and rural Britain. *Journal of Rural Studies* 5, 1–16.

Blowers, A. (1987) Transition or transformation? Environmental policy under Thatcher. *Public Administration* 65, 277–94.

Blunden, J. and Curry, N. (1988) *A Future for Our Countryside?* Basil Blackwell, Oxford.

Bollom, C. (1978) *Attitudes and Second Homes in Rural Wales*. University of Wales Press, Cardiff.

Capstick, M. (1987) *Housing Dilemmas in the Lake District*. Centre for North-West Regional Studies, University of Lancaster, Lancaster.

Clark, D. (1981) *Rural Housing Initiatives*. Paper 4: Self-building Schemes in the Countryside. National Council of Voluntary Organisations, London.

Clark, D. (1988) *Affordable Houses in the Countryside: a Role for Private Builders.* ACRE and the House Builders' Federation, London.

Clark, G. (1982a) Housing policy in the Lake District. *Transactions of the Institute of British Geographers*, NS 7, 59–70.

Clark, G. (1982b) *Housing and Planning in the Countryside.* John Wiley, Chichester.

Cloke, P.J. (1989) Rural geography and political economy. In: Peet, R. and Thrift, N.J. (eds) *New Models of Geography: the Political Economy Perspective, Volume 1.* Unwin Hyman, London, pp. 164–97.

Clout, H.D. (1969) Second homes in France. *Journal of the Town Planning Institute* 55, 440–3.

Clout, H.D. (1971) Second homes in the Auvergne. *Geographical Review* 61, 530–53.

Conservative Policy Centre (CPC) (1989) *Low-cost Housing in Rural Areas: A Conservative Strategy.* CPC, London.

Department of the Environment (DoE) (1977) *Housing Policy, a Consultative Document.* HMSO, London, Cmnd. 6851.

Department of the Environment (DoE) (1987) *Development Involving Agricultural Land.* Departmental Draft Circular, DoE, London.

EOTECH Research and Consulting Ltd (1989) *An Evaluation of the Craft Homes Experiment: Report to the Rural Development Commission.* Rural Development Commission Rural Research Series, No. 4, RDC, London.

Foulis, M. (1985) *Council House Sales in Scotland, April 1979 to December 1983.* Scottish Office Central Research Unit, Edinburgh.

Gibb, A. and MacLennan, D. (1986) Policy and process in Scottish housing, 1950 to 1980. In: Saville, R. (ed.) *The Economic Development of Modern Scotland, 1950–1980.* John Donald, Edinburgh, pp. 270–91.

Gilg, A.W. (1985) *An Introduction to Rural Geography.* Edward Arnold, London.

Grant, R. (1991) Second homes – a burning issue. In: Lewis, G.J. and Sherwood, K. (eds) *Rural Mobility and Housing.* Department of Geography, University of Leicester, pp. 17–38.

Herington, J. (1984) *The Outer City.* Harper and Row, London.

Hoggart, K. and Buller, H. (1987) *Rural Development – a Geographical Perspective.* Croom Helm, London.

Lake District Special Planning Board (LDSPB) (1980) *Cumbria and Lake District Joint Structure Plan, Written Statement.* LDSPB and Cumbria County Council, Kendal and Carlisle.

Lewis, G.J. and Bowler, I.R. (1987) The decline of private rented housing in rural areas: a case study of villages in Northamptonshire. In: Lockhart, D.G. and Ilbery, B.W. (eds) *The Future of the British Countryside.* Geo Books, Norwich, pp. 115–36.

McDowell, L. (1982) Urban housing markets (Unit 12, Block 3). In: Open University, *Course D 202 – Urban Change and Conflict.* Open University Press, Milton Keynes.

MacGregor, B.D., Robertson, D.S. and Shucksmith, M. (eds) (1987) *Rural Housing in Scotland: Recent Research and Policy.* Aberdeen University Press, Aberdeen.

Malpass, P. and Murie, A. (1987) *Housing Policy and Practice*. 2nd edn. Macmillan, London.

Massey, D.B. (1977) The analysis of capitalist land-ownership: an investigation of the case of Great Britain. *International Journal of Urban and Regional Research* 1, 404–24.

Massey, D.B. and Catalano, A. (1977) *Capital and Land: Land Ownership by Capital in Great Britain*. Edward Arnold, London.

Ministry of Housing (1944) *The Control of Land Use*. HMSO, London.

Ministry of Housing (1948) *The Design of Dwellings*. HMSO, London.

Mooney, E. (1991) Access to housing in rural Scotland: some preliminary observations in Bute. In: Lewis, G.J. and Sherwood, K. (eds) *Rural Mobility and Housing*. Department of Geography, University of Leicester, pp. 57–78.

Munton, R.J.C. and Goodchild, B. (1985) *Development and the Landowner*. Allen and Unwin, London.

National Agricultural Centre Rural Trust (NACRT) (1987) *Village Homes for Village People*. NACRT, London.

Newby, H. (1980) *Green and Pleasant Land? Social Change in Rural England*. Penguin, Harmondsworth.

Newby, H., Bell, C., Rose, D. and Saunders, P. (1978) *Property, Paternalism and Power: Class and Control in Rural England*. Hutchinson, London.

Pacione, M. (1979) Second homes on Arran. *Norsk Geografisk Tidsskrift* 33, 33–8.

Phillips, D.R. and Williams, A.M. (1981) Council house sales and village life. *New Society* 58, 367–8.

Phillips, D.R. and Williams, A.M. (1983) The social implications of rural housing policy. *Countryside Planning Yearbook* 4, 77–102.

Pratt, G. (1982) Class analysis and urban domestic property: a critical reexamination. *International Journal of Urban and Regional Research* 6, 481–502.

Reade, E. (1987) *British Town and Country Planning*. Open University Press, Milton Keynes.

Robinson, G.M. (1990) *Conflict and Change in the Countryside: Rural Society, Economy and Planning in the Developed World*. Belhaven Press, London and New York.

Robinson, G.M. (1991) EC agricultural policy and the environment: land use implications in the UK *Land Use Policy* 8, 95–107.

Rogers, A.W. (1983) Housing. In Pacione, M. (ed.) *Progress in Rural Geography*. Croom Helm, Beckenham, pp. 106–29.

Rogers, A.W. (1985a) Local claims on rural housing. *Town Planning Review* 56, 367–80.

Rogers, A.W. (1985b) Rural housing: an issue in search of a focus. *Journal of Agricultural Economics* 36, 87–9.

Rogers, A.W. (1987) Voluntarism, self-help and rural community development: some current approaches. *Journal of Rural Studies* 3, 353–60.

Rural Development Unit, Scottish Development Agency (SDA) (1989) *Rural Development in Scotland*. SDA, Edinburgh.

Sherwood, K. (1991) The local authority and local needs housing: a case study from South Northamptonshire. In Lewis, G.J. and Sherwood, K. (eds) *Rural*

Mobility and Housing. Department of Geography, University of Leicester, pp. 95–112.

Short, J.R., Witt, S. and Fleming, S. (1987) Conflict and compromise in the built environment: housebuilding in central Berkshire. *Transactions of the Institute of British Geographers*, NS 12, 29–42.

Shucksmith, M. (1981) *No Homes for Locals?* Gower, Farnborough.

Shucksmith, M. (1983) Second homes: a framework for policy. *Town Planning Review* 54, 174–93.

Shucksmith, M. (1984) *Scotland's Rural Housing: A Forgotten Problem* Rural Forum, Perth.

Shucksmith, M. (1987) Rural housing in Scotland: the policy context. In: Mac-Gregor, B.D., Robertson, D.S. and Shucksmith, M. (eds) *Rural Housing in Scotland: Recent Research and Policy.* Aberdeen University Press, Aberdeen, pp. 17–27.

Shucksmith, M. (1988a) Policy aspects of housebuilding on farmland in Britain. *Land Development Studies* 5, 129–38.

Shucksmith, M. (1988b) Trends in rural housing provision in Scotland. In Martin, T. and Doherty, J. (eds) *The Nature of the Scottish Housing Crisis.* Royal Scottish Geographical Society, Edinburgh, pp. 22–37.

Shucksmith, M. (1990) *Housebuilding in Britain's Countryside.* Routledge, London and New York.

Shucksmith, M. and Watkins, L. (1989) The development of housing associations in rural Scotland. A report to Scottish Homes. *Scottish Homes Research Reports* No. 1., Edinburgh.

Thrift, N.J. (1987) Manufacturing rural geography? *Journal of Rural Studies* 3, 77–81.

Whyte, D. (1978) Have second homes gone into hibernation? *New Society* 45, 286–8.

Wilcox, R. (1991) Rural housing and the role of new settlements: a case study of the New Plan for Kettering Borough. In: Lewis, G. J. and Sherwood, K. (eds) *Rural Mobility and Housing.* Department of Geography, University of Leicester, pp. 115–31.

Williams, N. and Sewel, J. (1987) Council house sales in rural areas. In: Mac-Gregor, B.D., Robertson, D.S. and Shucksmith, M. (eds) *Rural Housing in Scotland: Recent Research and Policy.* Aberdeen University Press, Aberdeen, pp. 66–85.

RECREATION AND TOURISM IN RURAL AREAS

8

DEFINING AND PROTECTING RURAL ENVIRONMENTS IN THE US

Robert Mason

America's settlement history is replete with contrasting and competing visions of 'urban' and 'rural' (Marx, 1964; Schmitt, 1969). Although rural sociologists, geographers and other scholars have recently called into question such social and spatial constructions (Cloke, 1989; Friedland, 1982; Gilbert, 1982; Hoggart, 1990), most acknowledge their value as research variables – albeit far from perfect ones. Ruralism – however it is constructed and manipulated – holds great popular and commercial appeal: ecologically whole, economically productive, and aesthetically pleasing rural landscapes bring satisfaction to consumers of the countryside (Ward, 1988). Indeed, the rural ideal has even found expression as a marketing concept for urban real-estate developers (Perkins, 1989; Stokes, *et al.*, 1989).

How do guardians of rural landscapes and environments decide what is worth protecting? What means do they employ for doing so? To the former, no simple answers emerge. Bound up in the explanations are such things as economic, racial and social segregation; protection of critical environmental zones; conservation of natural resources; maintenance and enhancement of property values; exclusion of unwanted facilities and land uses; historic preservation; and nostalgic longings. But from these diverse perspectives comes considerable agreement on the need to protect and preserve our coveted 'rural landscapes' (Stokes *et al.*, 1989).

Despite staunch resistance in many rural areas, zoning has been and continues to be the pre-eminent form of local land-use regulation in the US. In recent decades, however, various federal and state laws and regulations have sought to provide more central direction to local land-use regulation. By 1979, the great majority of states had proffered statutory protection to wetlands, floodplains, coastal areas, agricultural land and endangered species (Council on Environmental Quality, 1979). Federal air, water, coastal zone, agricultural preservation and other environmental legislation also came on line, principally during the 1970s. In varying degrees, these laws and regulations promote socioeconomic sorting, a process that permits and encourages expansion in some places and exclusion in others (Plotkin, 1987). Rather than attempt to describe the multitude of laws and policies that affect private land-use decisions, this chapter concentrates on the specific roles of greenline parks, greenways, and private land trusts in protecting rural environments.

Greenline Parks and Greenways

In its 1987 report, the President's Commission on Americans Outdoors (1987, p. 142) recommended that a nationwide network of private, local, and state 'greenways' be established to 'link together the open spaces of the American landscape'. Most greenways are linear in form, defined by such features as stream valleys, ridgelines, rail rights-of-way and canal towpaths (Didato, 1990; Little, 1990). The commission report also makes reference to protection of larger 'regional' landscapes, such as New Jersey's Pinelands National Reserve. As with greenways, the objectives are provision of recreation, and protection of habits and ecosystems. What distinguishes greenways and greenline parks from most conventional preserves in the US is that they include substantial proportions of privately owned lands. Greenline parks include significant areas of such lands within their boundaries; greenways transect private lands. Thus private lands, together with farms, homes, towns and economies, become part of the recreational and aesthetic experience. In this respect they are similar to national parks in the UK.

Conceptual evolution

Rather than explicitly calling for additional federal regulatory or oversight authority, the greenline concept favours intergovernmental ventures employing such land management techniques as tax incentives, special zoning plans, conservation easements, and purchase and lease-back arrangements (Corbett, 1983). The original greenline concept, like the later recommendations of the President's Commission on Americans Outdoors (1987), is rooted in the notion that recreation opportunities near urban areas are inadequate. Yet most articulations of the concept go beyond simple provision of recreation space. They call for protection of 'living landscapes'; those, that is, where people live and work. Rather than trying to move people out of the parks, or waiting for them to diminish in number by attrition, residents are encouraged to use land in 'appropriate' ways: farming, forestry, fishing, flood control, hunting and trapping, and carefully controlled tourism and housing development. Thus economic productivity coexists with environmental protection and provision of open space (Corbett, 1983; Little, 1983; Fogleman *et al*, 1985; Hirner, 1985; Hiss, 1990).

There is no formal system of greenline parks, and Congressional attempts to establish such a system faltered in the late 1970s. Yet several parks and recreation areas, to varying degrees, satisfy the general greenline criteria outlined above. Fig. 8.1 shows one version of the national greenline map, but different greenline proponents would draw different maps. Indeed, Little (1983) maintains that the only true greenline parks are New York State's Adirondack Park and New Jersey's Pinelands National Reserve.

The President's Commission on Americans Outdoors (1987), Little (1990) and other recreation advocates (Grove, 1990) have recently turned their energies toward 'greenways'. Greenways are generally smaller and more manageable than greenline parks. Although greenway designations generate conflicts – sometimes intense ones – usually they are neither so widespread nor prolonged as those associated with larger parks, such as the Pinelands (Mason, 1992) or Adirondacks (Heiman, 1988; Mason, 1991b). For these reasons, the prospects for success now seem greater for greenways than for greenline parks.

During the 1988 presidential campaign, George Bush lent his support to the idea of a trust fund that would generate 1 billion dollars per year for state and local land acquisition and development of recreational facilities (Bush later reneged). The fund would succeed the Land and Water Conservation Fund (LWCF), established for broadly similar purposes in 1966. Most LWCF funds, generated from offshore oil leasing revenues, remained unspent during the 1980s and early 1990s. Congressional proposals for an American Heritage Trust Act, which would create the new trust fund, remain tied up in subcommittee as of late 1991.

Both greenline parks and greenways carry significant implications for the aesthetic, economic and social restructuring of rural areas. A national programme with dedicated funding, as well as similarly structured state programmes, would give increasingly centralized direction to rural landscape evolution. Embodied in much of this pro-green sentiment are two important impulses: the desire to sustain a rural idyll (McLaughlin, 1986), and the perceived need to protect habitats and ecosystems.

Greenline and greenway examples

Heiman (1988, 1989) describes two New York State examples: the Adirondack Park and the Hudson River Valley. The historic Hudson River Valley has a long legacy of intensive productive as well as consumptive uses, and protection of its scenic views and historic estates has been championed by an impressive roster of government and industry elites. Regional planning efforts under Governor Nelson Rockefeller are described by Heiman

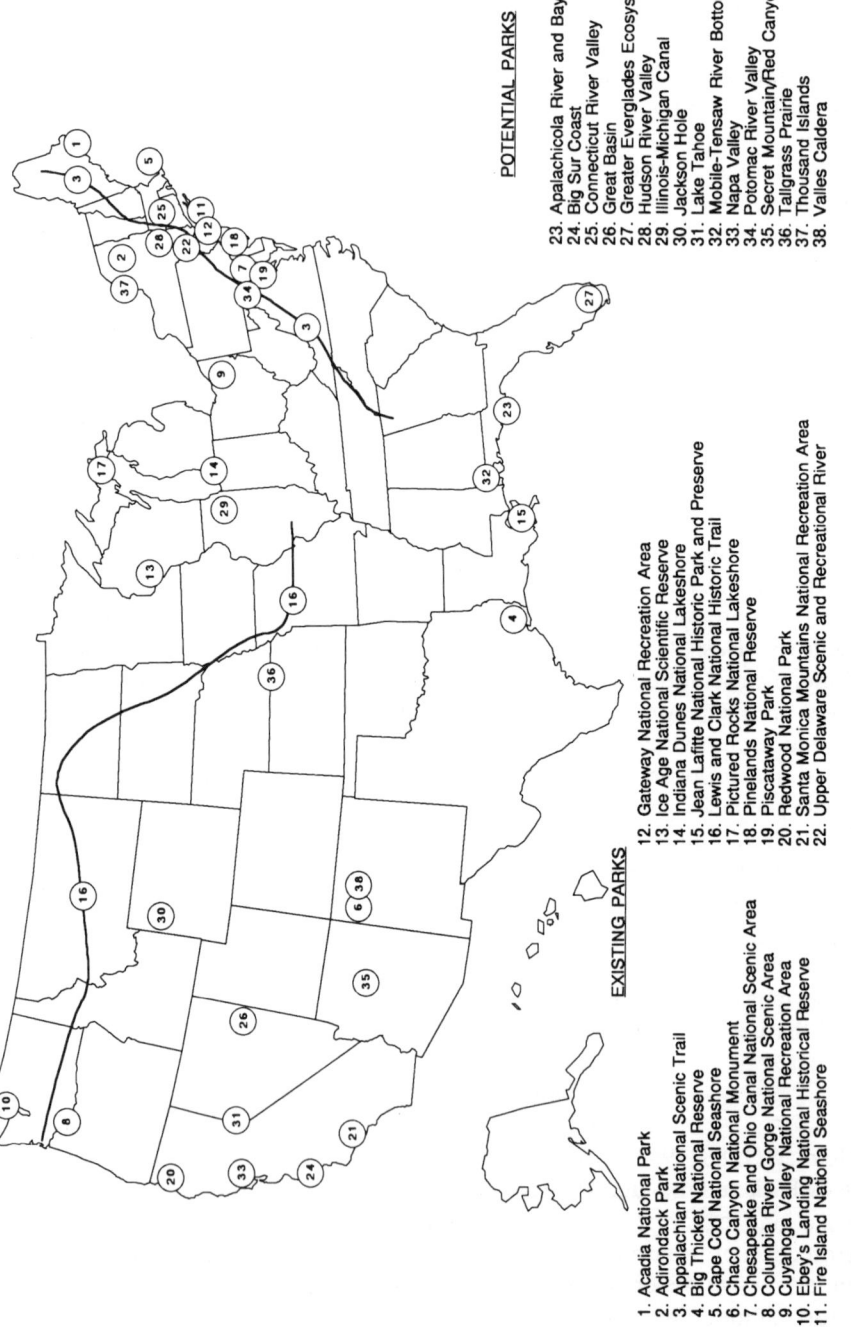

Fig. 8.1. Greenline parks. Sources: based principally on Corbett (1983) and Hirner (1985).

EXISTING PARKS

1. Acadia National Park
2. Adirondack Park
3. Appalachian National Scenic Trail
4. Big Thicket National Reserve
5. Cape Cod National Seashore
6. Chaco Canyon National Monument
7. Chesapeake and Ohio Canal National Scenic Area
8. Columbia River Gorge National Scenic Area
9. Cuyahoga Valley National Recreation Area
10. Ebey's Landing National Historical Reserve
11. Fire Island National Seashore
12. Gateway National Recreation Area
13. Ice Age National Scientific Reserve
14. Indiana Dunes National Lakeshore
15. Jean Lafitte National Historic Park and Preserve
16. Lewis and Clark National Historic Trail
17. Pictured Rocks National Lakeshore
18. Pinelands National Reserve
19. Piscataway Park
20. Redwood National Park
21. Santa Monica Mountains National Recreation Area
22. Upper Delaware Scenic and Recreational River

POTENTIAL PARKS

23. Apalachicola River and Bay
24. Big Sur Coast
25. Connecticut River Valley
26. Great Basin
27. Greater Everglades Ecosystem
28. Hudson River Valley
29. Illinois-Michigan Canal
30. Jackson Hole
31. Lake Tahoe
32. Mobile-Tensaw River Bottomlands
33. Napa Valley
34. Potomac River Valley
35. Secret Mountain/Red Canyons
36. Tallgrass Prairie
37. Thousand Islands
38. Valles Caldera

(1988, p. 238) as seeking to protect a 'privileged life-style from the ravage of poorly planned urban development, while also ensuring that public investment in energy, transportation, and residential development required for continued capital accumulation could continue'. More recently, the Hudson River Valley has become one of the nation's leading greenways. The avowed goal is coordinated regional growth, development and preservation. Because the greenway corridor is rather wide (and 150 miles long), and has within its borders 82 municipalities and counties, the Hudson River Greenway resembles a larger greenline-type park.

Further north, proposals for increased state regulation of the Adirondack Park have prompted a great deal of controversy. The Adirondack Forest Preserve was created in 1885, a comprehensive zoning and regulatory plan was put into place in 1973 and, in 1990, proposals were put forward for even more stringent land-use regulation (Mason, 1991b). Included with those proposals were recommendations for affordable housing programmes, incentives and subsidies for economic development, state-sponsored health care and education programmes, and more state aid to local government. Still, these sweeteners did little to assuage the anger of private landowners concerned about property taxes and local economies (approximately 42% of the Adirondack Park is in private ownership). As of late 1991, no new legislation had emerged.

Another arena for conflict is New England. In many minds, New England is equated not with its large cities, but with rustic rural ambience. Economic expansion and non-metropolitan population growth of the 1980s seriously threatened that image, prompting various calls for regional landscape protection (Lincoln Institute of Land Policy, 1986; Quinlan, 1987). A survey of New England's education, business and government leaders (New England Center, 1989) yielded the predictable calls for improved education, maintenance of a strong regional economy, and improvement of regional infrastructure. But the leaders also concurred on the need for conservation of agricultural land and natural environments.

This limited evidence does not translate into a conspiratorial reconstruction of the New England landscape by regional elites. But it does suggest that, at least in a climate of economic expansion, they are likely to support measures that seek to sustain the rural idyll. Landscape amenities are important to the health of the regional workforce and to the region's image. When in the late 1980s land speculators bought 40 500 ha of forest land in New Hampshire and Maine, and 38 850 ha in the Adirondacks, conservationists and elected officials became alarmed, and acted to reclaim at least parts of those acquisitions. A more concerted, proactive response came in the form of the Northern Forest Lands Study (Harper *et al.*, 1990), conducted under the auspices of the US Forest Service.

The study proposed various strategies for maintaining traditional land ownership and use patterns – in essence, a mix of large private holdings and publicly owned areas, with extensive public access throughout. Key among the study's recommendations are more state and local land-use controls, federal tax and spending incentives for conservation, federal and state land acquisitions, purchase of conservation easements, new federal funds for state and local land protection, and incentives for 'community improvement'. The northern forest area would in essence become a large greenline park. Although the regional leaders cited above – with the exception of some politicians and forest industry executives – are not visibly lined up in support of these proposals, the recommendations do, in part, reflect their interests. As was the case in the Adirondacks, though, a weakened economy has rendered vigorous new land-use controls improbable in the near term.

Equity considerations

Whose interests are served by greenline parks and greenways? What limited evidence there is points to middle- and upper-class recreationists, frequently to the exclusion of those less privileged. In their survey of urban greenway users, Furuseth and Altman (1991) found that users were predominantly well-educated, high-income, environmentally aware, young and white. In a broader context, Foresta (1984) contends that provision of mass recreation often conflicts with preservationist desires of the relative few who live in or take special pleasures from highly valued places. Foresta points out that planning for the Cuyahoga Valley (Ohio) and Santa Monica Mountains (California) National Recreation Areas, as well as Fire Island National Seashore (New York), has favoured preservationist interests over visitor access and recreation provision (see also Mitchell, 1978). As Donnelly (1987) points out with regard to national parks in Canada, England and Wales, use restrictions weigh most heavily on the working- and middle-class masses; upper middle-class and upper-class users – most of them landowners – are able to avoid restrictions because they have access to private lands where they can engage in 'forbidden activities'.

Environmental interest groups, ironically enough, have supported specific greenline-type proposals, but have not been strong advocates of greenline parks as a matter of policy. In general, they favour rigorous land-use restrictions, combined with maximum feasible public acquisition of environmentally sensitive lands. Their intransigence does, however, show signs of softening, especially with respect to the smaller-scale greenways.

Land Trusts

Land trusts are private, non-profit organizations – local, regional and occasionally multistate or national in scale – whose purposes are to secure protection of specific lands for purposes of ecological preservation, open space provision, recreation and/or other future uses. Land trusts may receive donations of land or easements, purchase land or easements, or act as facilitator, broker or negotiating agent between other parties. Land trusts often collaborate with governments, in many cases buying lands or easements for resale to government agencies. Donors or sellers of lands benefit from potential reductions in estate or gift taxes, lowered tax assessments, and favourable tax treatment accorded the charitable contribution itself.

The largest land trusts operate at the national level and are few in number. The Nature Conservancy claims to protect over 1.2 million ha of land; the Trust for Public Land about 0.96 million ha. As of 1991, 889 local land trusts belonged to the Land Trust Alliance. The Alliance collects data on the vast majority of local trusts which, through a combination of easements, land ownership, and conveyance of lands to other parties, are collectively responsible for protection of approximately 1.09 million ha of land (Land Trust Alliance, 1991). Most of the activity is in the New England, Middle Atlantic, West Coast and Great Lakes regions (Fig. 8.2). Recent growth in the number of land trusts and protected areas has been explosive.

These area figures seem rather trifling in comparison with the country's nearly 36 million ha of national and state parks. Yet land trusts act as more than mere 'gap stoppers' in their efforts to protect valued lands. Indeed, Tunbridge (1981) sees Britain's National Trust, whose presence is more pervasive than that of any US trust, as an important agent in shaping landscapes, land uses and rural economy and society. Rose (1988) argues that land trusts have perpetuated elitist ideology – based on artist Andrew Wyeth's landscape images – in the Chadd's Ford area of southeastern Pennsylvania.

Land trust influence is broadened in several ways:

1. Land trusts may acquire interests in strategically situated parcels of land, thus blocking or inhibiting development plans over much larger areas;
2. Although their tax-exempt status limits them from engaging in overtly political activities, land trusts often work closely with legislators and government officials, and many enjoy wide political support;
3. Some land trusts offer services to other land trusts, and to local governments and planning agencies;
4. Regional land-trust consortia are increasing in number, and regional

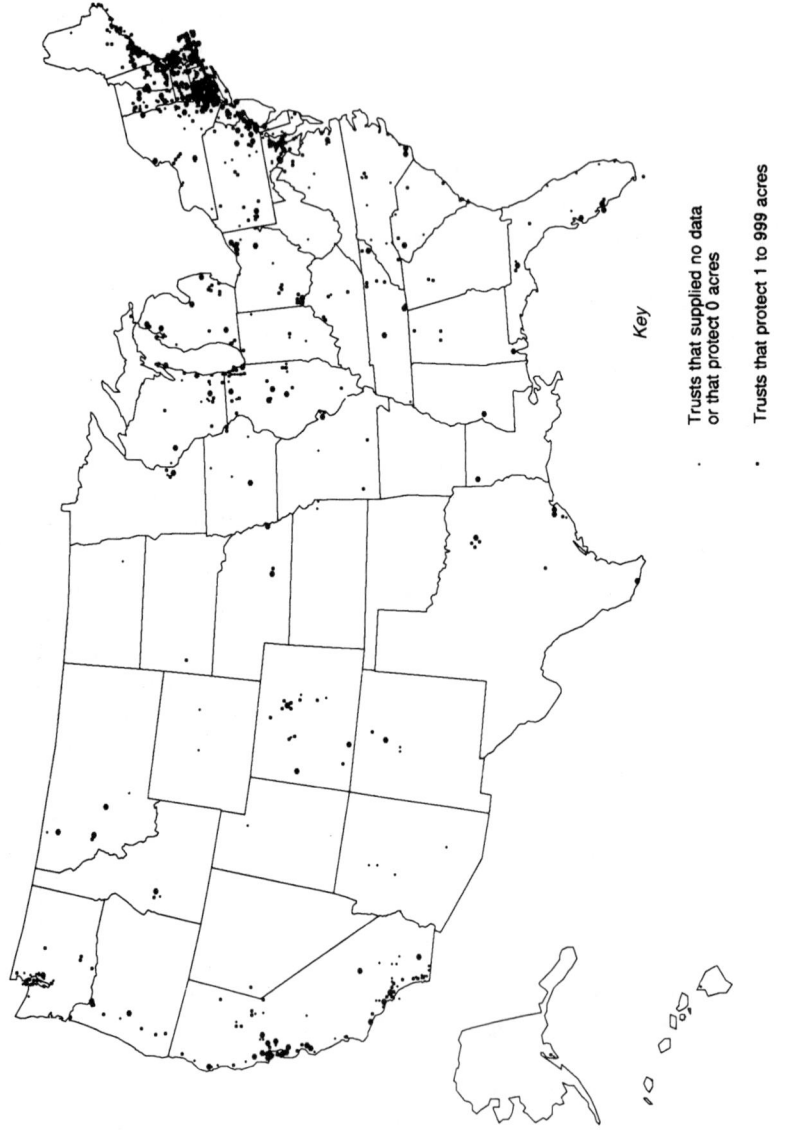

Key

· Trusts that supplied no data
  or that protect 0 acres

· Trusts that protect 1 to 999 acres

• Trusts that protect 1000 or more acres

Fig. 8.2. Local land trusts. Source: Land Trust Alliance (1991).

coordination was a theme of the Land Trust Alliance's 1991 national conference;

5. Most land trusts engage in public education activities;

6. Direct infusion of state or federal funds allows land trusts to expand their activities and broaden their influence. As currently proposed, the American Heritage Trust Act would ensure that at least 5% of all funds dispersed would go to private non-profit organizations.

Although ecological protection is the foremost concern of a majority of local land trusts (Land Trust Alliance, 1991; Mason, 1991a), large numbers of trusts are concerned with protecting rural landscapes. Mission statements and promotional literature commonly reflect this orientation, as do survey results nationally (Land Trust Alliance, 1991) and in the states of Massachusetts, Pennsylvania and California (Mason, 1991a). While this shared vision of the ideal rural landscape may be elusive and blurred, it is nonetheless compelling.

Conclusion

Much of the activity described above is taking place at the urban and suburban fringes of America's megalopoli. Does this portend major transformations in the ways we manage economies and land use in exurbanizing America? Will our interventions create landscapes of consumption suited to politically influential professional and managerial classes? There is evidence that this, indeed, will increasingly be the case in at least some highly valued regions, such as the Hudson Valley and New England. Nationally, this orientation is reflected in the 1991 Congressional hearings on the 'Impact of Transportation Policy on the American Landscape', as well as in the proliferation of national organizations – among them Open Space America, Successful Communities, and the Conservation Fund – that promote the cause through educational efforts and by funding specific projects. The recently inaugurated *Countryside* magazine earmarks a portion of its revenues for private land conservation organizations, and the Hudson River Valley's *UpRiver/DownRiver* may be the first magazine devoted exclusively to protection of a regional amenity landscape.

One of the likely consequences of more methodical regional landscape protection is a reinforced pattern of socioeconomic segregation. Some programmes speak to this concern by seeking to provide affordable housing. Vermont's state-subsidized Housing and Conservation Trust Fund is an important example; the New Jersey Supreme Court's Mount Laurel decisions, which require municipalities to provide for affordable housing, constitute another. More often, though, exclusion would seem to be the norm. At a recent gathering celebrating 10 years of regional planning in

New Jersey's Pinelands, land-use economist James Nicholas argued that transferable development rights produce affordable housing. By increasing density requirements and associated economies of scale in designated areas, they leave other prized areas open for the more exclusive occupance of those who can afford to buy enough land to live there. Nicholas' view (though not my interpretation of it) appeals widely to environmentalists, land-use planners, politically influential residents of exurban areas, and at least a part of the development community.

The orderly and aesthetically 'correct' landscape planning, of which planners' dreams are made, may never prevail across the country's great expanse, but it will probably continue to find increased favour in certain areas. Rural areas that attract recreational development, as well as other service and high-technology enterprises, might best fit the bill. Many such places, as exemplified by greenline park and land trust activities (Figs 8.1 and 8.2) are likely to be close to heavily populated regions. Where cherished rural images are well-ensconced in the general consciousness, increased planning activity is all the more probable.

What is being described, in large part, are those places where expansion gives way to exclusion. Not coincidentally, as in the Adirondacks and northern New England, local populations fiercely resist such landscape transformations, or at least the institutions and rules that would guide landscape change. Paradoxically, some valued rural regions will be prevailed upon to balance amenity with the residuals of production: airports, transmission lines, electric generating stations, low-level nuclear waste repositories, and other such intrusions. Powerful alliances that seek to define and sustain the rural idyll may, as a consequence, fracture and recrystallize in unorthodox and unpredictable ways. With a fair degree of certainty, it is possible to specify many of the places likely to be strongly contested. More elusive, though, will be the nature and timing of specific rural landscape protection efforts.

References

Cloke, P. (1989) Rural geography and political economy. In: Peet, R. and Thrift, N. (eds) *New Models in Geography, Vol. 1.* Unwin Hyman, London, pp. 164–97.

Corbett, M.R. (ed.) (1983) *Greenline Parks: Land Conservation Trends for the Eighties and Beyond.* National Parks and Conservation Association, Washington DC.

Council on Environmental Quality (1979) *Environmental Quality 1979.* US Government Printing Office, Washington DC.

Didato, B. (1990) The paths less travelled: a wrapup on the nation's Greenways. *Planning* 56, 6–10.

Donnelly, P. (1987) Creating national parks – 'a grand, good thing?' *Tourism Management* 8, 349–51.

Fogleman, V., Hirner, D.K. and Mertes, J.D. (1985) Greenline parks: an Anglo-American comparison with emphasis on national recreation areas, national seashores and national recreational and scenic rivers in the United States. Unpublished manuscript.

Foresta, R.A. (1984) *America's National Parks and Their Keepers*. Resources for the Future, Washington DC.

Friedland, W.H. (1982) The end of rural society and the future of rural sociology. *Rural Sociology* 47, 589–608.

Furuseth, O.J. and Altman, R.E. (1991) Who's on the greenway: socioeconomic, demographic, and locational characteristics of greenway users. *Environmental Management 15*, 329–36.

Gilbert, J. (1982) Rural theory: the grounding of rural sociology. *Rural Sociology* 47, 609–33.

Grove, N. (1990) Greenways: paths to the future. *National Geographic* 177, 77–99.

Harper, S.C., Falk, L.L. and Rankin, E.W. (1990) *The Northern Forest Lands Study of New England and New York*. US Department of Agriculture, Forest Service, Rutland, Vermont.

Heiman, M.K. (1988) *The Quiet Evolution: Power, Planning, and Profits in New York State*. Praeger, New York.

Heiman, M.K. (1989) Production confronts consumption: landscape perception and social conflict in the Hudson Valley. *Environment and Planning D: Society and Space* 7, 165–78.

Hirner, D.K. (1985) Public parks on private lands: Greenline parks protecting landscapes of national significance. PhD dissertation, Texas Tech University.

Hiss, T. (1990) *The Experience of Place*. Knopf, New York.

Hoggart, K. (1990) Let's do away with rural. *Journal of Rural Studies* 6, 245–57.

Land Trust Alliance (1991) *1991–92 National Directory of Conservation Land Trusts*. Land Trust Alliance, Washington DC.

Lincoln Institute of Land Policy (1986) *A Land Conservation Strategy for the New England Region*. Lincoln Institute of Land Policy, Cambridge, Massachusetts.

Little, C.E. (1983) The national perspective: greenline parks. In: New York State Department of Environmental Conservation, *Proceedings: Greenline and Urbanline Parks Conference*, DoEC, New York, pp. 3–5.

Little, C.E. (1990) *Greenways for America*. Johns Hopkins University Press, Baltimore.

Marx, L. (1964) *The Machine in the Garden: Technology and the Pastoral Ideal in America*. Oxford University Press, New York.

Mason, R.J. (1991a) Land trusts as shapers of rural landscapes. Paper presented at annual meeting of Association of American Geographers, Miami.

Mason, R.J. (1991b) Local responses to proposed new land use controls in New York's Adirondack Park. In: Hanink, D.M., Hovinen, E. and Hovinen, G. (eds), *Proceedings, Joint Meeting of the New England-St. Lawrence Valley Geographical Society and the Middle States Division of the Association of American Geographers*, Norwich, Conneticut, pp. 7–15.

Mason, R.J. (1992) *Contested Lands: Conflict and Compromise in New Jersey's Pine Barrens*. Temple University Press, Philadelphia.

McLaughlin, B. (1986) Rural policy in the 1980s: the revival of the rural idyll. *Journal of Rural Studies* 2, 81–90.

Mitchell, J.G. (1978) The re-greening of urban America. *Audubon* 80, 29–52.

New England Center (1989) *The Future of New England Project: Final Report*. New England Center, Durham, New Hampshire.

Perkins, H.C. (1989) The country in the town: the role of real estate developers in the construction of the meaning of place. *Journal of Rural Studies* 5, 61–74.

Plotkin, S. (1987) *Keep Out: The Struggle for Land Use Control*. University of California Press, Berkeley.

President's Commission on Americans Outdoors (1987) *Americans Outdoors: The Legacy, the Challenge*. Island Press, Washington DC.

Quinlan, R. A. (ed.) (1987) *Planning for the Changing Rural Landscape of New England: Blending Theory and Practice*. New England Center, Durham, New Hampshire.

Rose, D. (1988) The Brandywine: a case study of an ecological strategy. *Landscape Journal* 7, 128–33.

Schmitt, P.J. (1969) *Back to Nature: The Arcadian Myth in Urban America*. Oxford University Press, New York.

Stokes, S.N., Watson, A.E., Keller, G.P. and Keller, J.T. (1989) *Saving America's Countryside: A Guide to Rural Conservation*. Johns Hopkins University Press, Baltimore.

Tunbridge, J.E. (1981) Conservation trusts as geographic agents: their impact upon landscape, townscape and land use. *Transactions of the Institute of British Geographers* N S 6, 103–25.

Ward, E.J. (ed.) (1988) *Aesthetics of the Rural Renaissance: Proceedings of the 1987 Conference*. California Polytechnic State University, San Luis Obispo, California.

9

RECREATION, ACCESS, AMENITY AND CONSERVATION IN THE UNITED KINGDOM: THE FAILURE OF INTEGRATION

Nigel Curry

Four Distinct Components of Rural Leisure

In both political pressures for change and the countryside law of England and Wales a distinction always has been made between: *recreation*, which concerns social well-being (and historically, public health) and opportunities for enjoyment; *access*, which relates to the legal basis of rights over land (usually private) and exceptions; *amenity* (or scenic or landscape) conservation, which concerns the aesthetic worth of the landscape; and *nature conservation*, concerning the scientific values of ecosystems.

The separation of recreation and access

Policies and legislation for recreation and access have developed in parallel but separately. Prior to the 1949 *National Parks and Access to the Countryside (NPAC) Act* several 'middle class' pressure groups such as the (then) Commons, Open Spaces and Footpaths Preservation Society and the Cyclists Touring Club, seeking active use of their increasing leisure time, were vocal during the 1930s. They pressed for the enhancement of open-air enjoyment, principally for health reasons.

The access lobby was more vociferous (Stephenson, 1989). The Ramblers' Federation, made up of 'working class' rambling clubs, was seeking an antidote to difficult working conditions. A mass trespass of Kinder Scout, during the 1930s, was part of an increasing frustration of the failure to improve access opportunities through Parliament. There had been eight 'Access to Mountains' Bills defeated between 1908 and 1939 by a Parliament dominated, particularly in the House of Lords, by a 'landowning class' (Dower, 1978).

In preparation for the 1949 Act, the Hobhouse Committee (1947), which was set up to report on the location of national parks 'for public enjoyment', required a separate subcommittee to examine the legal basis of access to private land (Ministry of Town and Country Planning, 1947). The Act itself contained distinct provisions for public enjoyment (national parks) and access (access agreements and the definitive map).

Preceding the 1968 *Countryside Act* too, a White Paper 'Leisure in the Countryside' (Ministry of Land and Natural Resources, 1966) set out comprehensive proposals for public enjoyment. The Gosling Committee (1967) considered rights of way separately. In the Act itself, clauses 1 to 10 relate to recreation, and 15–21 to access (Cherry, 1975).

Since the 1968 Act, there has been little legislation for recreation, excepting the 1986 *Agriculture Act* which considers it as a possible means of farm diversification. Access, conversely, has been addressed in the 1981 *Wildlife and Countryside Act* (the completion of definitive maps) and the 1990 *Rights of Way Act* (rights of way obstructions, ploughing footpaths).

Disaggregating amenity and scientific conservation

Pressures for the protection of the countryside had, by the 1930s, bifurcated into distinct amenity and scientific schools. The National Trust was championing aesthetic and amenity conservation, and the British Correlating Committee scientific and ecological conservation (Woods, 1984). Here too, Hobhouse took the amenity brief himself, twinning it with recreation, but set up a separate 'nature conservation' committee, chaired by Sir Julian Huxley. Nature conservation provisions were passed by both Houses of Parliament with little discussion, but amenity conservation measures were hotly debated.

Distinct conservation provisions came in the 1949 *NPAC Act*. National Parks and Areas of Outstanding National Beauty (AONBs), together with management agreements for specific landscape protection, were to serve amenity interests. These were to embrace access and public enjoyment. National Nature Reserves (NNRs) and Sites of Special Scientific Interest (SSSIs) were provisions for scientific conservation and were to be largely exclusionary. Only local nature reserves were to tolerate access. Even institutionally, the (then) National Parks Commission took the amenity role, and the (then) Nature Conservancy the scientific one.

By the 1968 *Countryside Act*, the newly-formed Countryside Commission was to have responsibility for 'the enhancement of natural beauty and amenity', but scientific conservation responsibilities were more ubiquitous – 'every minister, government department and public body shall have regard to the desirability of conserving fauna and flora' (HMSO, 1968). Few new provisions were produced for either in the Act.

The 1973 *Nature Conservancy Council Act* contained only scientific conservation provisions, separating research and advice. The 1981 *Wildlife and Countryside Act* (amended in 1985), however, introduced new-style management agreements for amenity landscape changes. The renotification of SSSIs and species protection provisions provided scientific counterparts. The 1986 *Agriculture Act* now extends the responsibility for amenity conservation to all government Ministers.

Thus these four areas, at least three of which are centrally concerned with rural leisure (even the fourth, scientific conservation, has an increasingly important role in this respect (McLaughlin and Singleton, 1979; Roome, 1982)), have been distinct in their legislative, policy and institutional evolution. This has often caused ineffectiveness in the implementation of policies, particularly for designated areas, and the compromising of objectives. In particular, both recreation and access objectives have been residualized and conservation ones given priority.

Relegating Recreation and Access

There was policy enthusiasm for recreation preceding 1949. Both Dower (1945), in examining the basis for national parks, and Hobhouse (1947), recommended enhanced 'recreation opportunities for the whole nation'. In fact, these opportunities were not to be for all. Dower felt that some people would be better off pursuing their interests in an urban setting and Hobhouse was keen to limit the extent of recreation. These sentiments found their way into the 1949 *NPAC Act*. Recreation was to be condoned if it was to be peaceful and did not disturb anyone and was without sacrifice to conservation goals. For access, both Dower and Hobhouse favoured the right to wander at will in open countryside and £50 million was earmarked prior to the Act to ease this. In introducing the 1949 Act, Lewis Silkin, Minister of Town Planning, said that it was to be:

> a people's charter for the open air, for the hikers and the ramblers, for everyone who loves to get out into the open and enjoy the countryside.
>
> (Blunden and Curry, 1990, p. 63)

In the Bill's passage, though, the landowning interests in Parliament had their effect. At the passing of the Act Silkin acceded:

> a person's land is his land. I think it is wrong to give the public an automatic right to go over all private land.
>
> (Blunden and Curry, 1990, p. 90)

The 1949 *NPAC Act*, and all Acts since, thus failed to achieve the free access to open country. By the mid 1950s, the £50 million set aside to ease access appeared to have evaporated.

Two conflicting notions characterized recreation policy during the 1950s and 1960s. There was a political will to make up for some of the failings of the 1949 Act but there was also an emerging fear of a recreation 'explosion'. The 1968 *Countryside Act* thus considered recreation as both an opportunity and a potential land-use problem. Country parks, picnic sites and transit caravan sites were introduced as recreation facilities rather than access rights, both for opportunities and to relieve pressures. Niall McDermott, in moving a second reading of the Bill in 1967 described it as a:

> comprehensive Bill designed to tackle the problems of the
> countryside – problems which are increasing at an increasing pace.
>
> (Curry, 1986, p. 13)

Both the 'Countryside in the 1970s' Conferences (Council for Nature, 1966) and the House of Lords Select Committee on Sport and Leisure (1973) cautioned against the recreation explosion. The resultant implementation of these recreation parts of the 1968 *Countryside Act*, chiefly through county councils and the structure planning process, was almost universally restrictive. An analysis of all structure plan policies to 1985 reveals the steerage of recreation away from 'high quality' landscapes towards areas of lesser amenity value, as well as the intention in the main to provide only passive recreation pursuits. The use of agricultural land for recreation was universally rejected. National park plans were also preoccupied with recreational containment (Curry and Comley, 1985).

There have been some changes in these attitudes in the late 1980s. Food surpluses and the 1986 *Agriculture Act* have prioritized public enjoyment, principally for its economic worth. Structure plan and national park plan reviews also are more acquiescent, preferring the notions of good management to damage limitation. The recreation 'explosion' had either been successfully resisted, or had never actually happened.

Since the 1949 *NPAC Act*, then, the recreation imperative has been thwarted by 'no threat to amenity' and 'fears of a recreation explosion' characteristics of policy. A tolerance of 'quiet enjoyment' has prevailed. Analysis of participation in this quiet enjoyment, however, has shown it to be skewed towards the more affluent (Curry and Comley, 1986). Recreation policies thus have not only been residualized, but distributionally regressive.

Of 1949 *NPAC Act* access policies, the recording of rights of way on a 'definitive map', to be completed within and subsequently reviewed every 5 years, was an unrealistic target. For recorded rights of way, maintenance has always been a problem, as noted by Glyptis (Chapter 10). Also few access agreements have ever been introduced and most lie in the Peak National Park. Many of these are due for renegotiation in 1993 and difficulties in successfully achieving this are feared. The provision for Long Distance Routes was acted upon, but the first was not opened until 16 years after the Act.

For common land, a 1958 Royal Commission recommended access for quiet enjoyment on foot for all commons. Still no statute provides for this, despite two attempts during the 1980s to introduce common land Bills into Parliament and agreement to the principle by the Common Land Forum (an amalgam of the access lobby and landowners) in 1986.

Thus, access provisions, too, have made little progress since the 1949 *NPAC Act* chiefly, it must be said, because of the strength of the landowning interest in policy-making and the reliance on voluntary effort in producing the definitive map. The Countryside Commission still espouses the virtues of access agreements (Countryside Commission, 1987) and the need further to develop rights of way (Countryside Commission, 1989a); but its role remains advisory to the shire counties where the political will often is not strong.

Prioritizing Amenity and Scientific Conservation

Whilst scientific conservation provisions in the 1949 *NPAC Act* were largely accepted by Parliament, amenity conservation provisions became its main preoccupation. This centred on whether planning powers to control landscape changes in designated areas should be nationally or locally based. This debate, which has never satisfactorily been resolved, ensured that the heritage of the Act was clearly a conservation rather than a recreation or access one (Cherry, 1985).

By the 1960s, fears of a recreation explosion elevated conservation objectives during the passage of the 1968 *Countryside Act*. Although the Bill was 'first and foremost about opportunities for the enjoyment of the countryside' (Department of the Environment, 1967), its conservation considerations were inserted during its Committee stage. Importantly, the Countryside Commission's powers were extended to cover 'the enhancement of natural beauty and amenity' (Department of the Environment, 1967). The Bill itself had contained only recreation and access functions for the Commission.

In the early 1970s, the failure of the development control system adequately to stem landscape deterioration in designated areas, particularly brought about by agriculture and forestry, led to the conservation priority becoming more formalized. The Sandford Report (1974), in reviewing the policies of national parks, proposed that where the two were in unavoidable conflict, conservation should take priority over recreation. By the 1980s, this had been given formal sanction in the Countryside Commission's prospectus (Countryside Commission, 1982, p. 4) on accession to 'grant in aid status' under the 1981 *Wildlife and Countryside*

Act, a significant shift from its intended role in 1967: 'proportionately rather more of our resources will go into conservation', although 'recreation and access will continue to receive a substantial part of our funds'. This priority of amenity and scientific conservation over recreation and access is also evident outside of national policy. Many more conservation designations have been introduced than recreation ones, for example, and structure plan conservation policies have been promotional (Penning-Rowsell, 1983; Bain *et al.*, 1990); but recreation policies have been dominantly restrictive. A similar picture pertains in local plans and countryside strategies (Curry, 1985, 1991). In national park plans too, the priority is clear. In the North York Moors National Park for example:

> it has never been part of the national park philosophy to match, automatically, recreational demands but rather temper these demands to the ability of the landscape to absorb them.

In all national parks during the 1980s, expenditure on conservation has exceeded that of recreation (Countryside Commission, 1990).

The increasing concern for amenity and scientific conservation objectives over the past 15 years or so has been legitimate. The failure of the planning system to stem adequately development pressures in national parks has been charted (Curry and Edwards, 1991) but it has been landscape change as a result of agriculture and forestry practices, largely outside of planning control, that has generated most concern. The ubiquitous nature of this over the whole of England and Wales (Shoard, 1980; Pye-Smith and Rose, 1984), and the response by pressure groups (Lowe and Goyder, 1983), has ensured an increasing political importance of such ravages to the landscape and ecology of the countryside.

The legitimacy of prioritizing conservation over recreation is less clear. Undoubtedly in specific, often celebrated cases, recreation damage to the environment is critical. There is no comprehensive evidence, however, to suggest that its extent is that great, and such research that exists suggests that it is not so widespread as is commonly presumed (Siddaway and O'Connor, 1978). Even where it is evident, positive and integrated management would perhaps offer a more satisfactory alternative than conservation prioritization.

Policies for Designated Recreation and Amenity Landscapes

Notwithstanding the relative importance of recreation, access, amenity and scientific conservation policies, it was in the designated landscapes of the 1949 *NPAC Act* that their implementation was to be critical. National parks were to consider amenity conservation and recreation and access.

AONBs were to focus on the first only, reinforcing the conservation priority.

National parks

Two principal concerns in the development of national parks were: who should have planning powers over amenity conservation, and the organization and resourcing of parks. Both stemmed from the private ownership of land in parks. The 1949 Act provided for joint planning boards for multicounty parks, independent of county councils but made up of constituent members. Single-county parks were to be planned by the counties, with a National Park Commission having only an advisory function.

Only two parks – the Lake District and the Peak District – were made independent joint boards. Through vociferous local objection, a joint advisory committee was accepted for Snowdonia, which set a precedent for all of the other parks designated during the 1950s. As a result, the three principal planning intentions for parks of the 1949 Act, that is, independent joint boards, a full-time planning officer and the planning of parks as one individual unit, were lost. There was thus very little 'national' distinctiveness about the parks. The 1972 *Local Government Act* reconstituted boards in the Peak District and the Lake District, but elsewhere required only that each national park should have a single county council committee, which was to appoint its own official, a national park officer, and produce a national park plan.

The 1972 Act introduced stronger planning powers but more importantly, through officers and plans, encouraged stronger policies addressing a concern of the Sandford Report (1974). The 1991 *Planning and Compensation Act* develops these policy strengths by requiring parks to produce park-wide local plans. The organizational structure of parks, however, has never been satisfactorily resolved.

A third problem for parks has been the amenity impacts of agriculture and forestry. No control for these was built into the 1949 *NPAC Act*, since none was then considered necessary. Sandford (1974) saw these sectors as the principal threat to the quality of parks, particularly in relation to moorland ploughing. This had reached a critical stage in some parks, notably Exmoor, by the mid 1970s.

The resultant report (Porchester, 1977) influenced provisions for the notification of such operations and the requirement to compensate, or make management agreements, which were introduced in the 1981 *Wildlife and Countryside Act*. The 1986 *Agriculture Act* also placed conservation duties on Agriculture Ministers, and parts of two Environmentally Sensitive Areas fall into national parks. Experiments to control agricultural and forestry impact are also taking place, but these piecemeal initiatives have provided no real substitute for comprehensive planning powers over the principal land users in parks.

AONBs

AONBs were to be smaller and less wild than national parks. Ironically, they would be commonly designated in lowland, more accessible areas of the country, but would have no specific access or recreation functions. AONB designations were slower than for parks: 19 had been introduced by the mid 1960s, 32 by the mid 1970s, and 38 by the designation of the North Pennines in 1988. They now cover 13% of the land area of England and Wales, compared with 10% for national parks.

During the 1960s, policies in AONBs had been a matter of slightly more stringent development control. The National Parks Commission had sought greater conservation powers for them in 1961 and, in response, the government actively considered, but rejected, their abolition. The Commission produced a policy statement at that time requesting the formation of joint advisory committees between the counties in which they fell and 'statements of intent' for their development. The response to these was, and remains, disappointing.

By 1978 the, by now, Countryside Commission produced an AONB policy consultation paper which again included the option of scrapping them altogether. This was followed by a report to the Commission (Himsworth, 1981) that recommended better planning and management for AONBs, the inclusion of recreation objectives, and the extension of agricultural operation notification procedures required in national parks to AONBs. These proposals were put by the Commission to the Department of the Environment, which rejected the second two.

The resultant Countryside Commission (1980) policy statement reaffirmed amenity conservation as the principal objective of AONBs, and championed the development of management plans and stricter development control in them. The establishment of joint advisory committees and statements of intent was again encouraged.

Overall AONBs have suffered from a lack of any special powers to pursue their amenity goals. Himsworth (1981) admitted that nothing had been done in AONBs that could not have been done under normal planning powers, whereas procedurally they certainly have much more in common with the undesignated countryside than with the national parks. Apart from a few joint advisory committees, they have no separate administration or finance, and in only a minority of cases are planning officers in local authorities given special AONB responsibilities. In the 1990s they suffer from increasing recreation pressures, and particularly landscape and habitat damage, with no immediate resolution in sight (Smart and Anderson, 1990).

Policy Reformulation to the Turn of the Century

This chapter has argued that the consideration of recreation, access, amenity and scientific conservation has been developed in an uncoordinated way and that the first two have been residualized relative to the other two. It has outlined enduring problems in the development of national parks and AONBs since designation. Is there likely to be any significant change in these characteristics over the next 10 years? Two recent policy reviews, for national parks and AONBs, help to inform this question. The National Parks Review Panel (Edwards, 1991) explicitly proposes reformulating national park purposes to give even more weight to conservation relative to 'public enjoyment' objectives.

Conservation objectives should now explicitly embrace scientific as well as amenity conservation, and recreation objectives should be more closely tied to the special qualities of parks. Sandford's (1974) priority – where the two are in conflict, conservation must have a clear priority over recreation – is reaffirmed. The report thus calls conservation the 'first' purpose of national parks, and recreation the 'second'.

This priority is borne more from the importance of conservation, in the face of damaging economic land uses, than any overwhelmingly detrimental impact of recreation. Certainly there is an intention to avoid environmentally intrusive leisure pursuits, particularly relating to motor sports, but the report does accede that: 'activities . . . are frequently focused on small areas. Such intensive local use can lead to erosion and other problems' (Edwards, 1991, p. 42). This suggests that recreation damage may not be universal in parks and the report calls for further research in this area. There are also fears of further growth in the number of visits to parks unless containment is developed; this is reminiscent of the fears of the recreation explosion in the 1960s and the 1970s.

For national parks, therefore, the residual nature of recreation objectives seems set to continue; this is despite conservation problems in parks being driven by agriculture, forestry and other developments rather than recreation, and the extent of recreation damage being unclear. Positive elements in the review include management plans and codes of practice, and only in the case of forestry operations is it proposed that recreation be integrated. The principal proposals for recreation focus on promoting it in areas surrounding parks, community forests and other areas, and generally 'taking the park to nearby towns and conurbations'. There is little change here in seeing recreation as a land-use problem and seeking to 'control' it in some way – in this case by shifting the burden to other areas.

In contrast, the review strongly defends the rights of access in parks. Full definitions of rights of way are required urgently, and comprehensive (though not universal) access to open country should be secured through existing mechanisms. Erosion problems should be tackled by positive management and the power to create new rights of way should be exploited. It is also recommended that the Common Land Forum's views should be brought to legislation.

Organizationally, the review proposes that all parks become independent 'national park authorities'; they should have more resources and increased planning powers to include joint park-wide local (development) plans and national park (management) plans, with a partnership input into structure plans.

The recent AONB policy review (Smart and Anderson, 1990) sustains the existing statutory amenity conservation priority over recreation and access. Perhaps predictably, it suggests that recreation 'pressures' (it is still largely seen as a problem) are most appropriately 'dealt with' through better education and interpretation. Recreation and access command little attention in the report overall, however. It does discuss the notion of a more holistic 'integrated' management approach to the management of land use, but is cautious of its practicalities.

Resourcing and organizational problems of AONBs are discussed at some length and, whilst making a number of recommendations here (such as jointly managed budgets, AONB officers, management plans, sponsorship and so on), the report remains pessimistic about their successful implementation. This is both because of a lack of compulsory powers on the part of the Countryside Commission and the residual status of AONBs relative to national parks.

For the longer term, greater policy effectiveness is proposed through the coercion of differential funding. The Countryside Commission would initially identify priority AONB areas for protection, with resultant targeted resourcing. But this selective funding would be conditional on the relevant AONBs having joint advisory committees, management plans, policies in countryside strategies, better monitoring of expenditure and a 'high standard of policymaking and management ability'. This is a novel approach to changing the historically low level of commitment to these, but could lead to the anomaly of the Commission failing to allocate targeted funds. It also could lead to a further residualization of non-targeted AONBs.

Thus, both of these reports have little new to say, save the detail, about the organization and resource structures of national parks and AONBs that has not already been proposed, and rejected, by governments in earlier policy reviews. Positive management measures to overcome recreation pressures and its integration with conservation are not fully discussed in either report. Where such integrated management is considered for AONBs, it is considered unrealistic. This is unfortunate, since some of the pioneering work on fully-integrated land management, such as the upland management experiments (Countryside Commission, 1977) and the Peak Park integrated rural development experiments (Peak District National Park Joint Planning Board, 1991), actually have taken place in national parks.

Conclusion

Ultimately the integration of recreation, access, amenity and scientific conservation policies must be the most successful means of considering these four components of rural leisure: as Benson (1986) notes, amenity and scientific conservation are essentially the supply of landscapes and ecosystems, and countryside recreation and access the demand for them. Prioritizing one over the other is likely to encourage a supply and demand imbalance.

Not only have reports for the 1990s failed successfully to integrate these four components, but there are three political factors that will inhibit the full adoption in government policy of what they have proposed. The first is resources. Even though more government funds are disposed to conservation than recreation and access, resources to both are still minuscule relative to those devoted to agriculture and forestry. For the AONB review (Smart and Anderson, 1990) to have to discuss alternative sponsorship strategies to resource AONBs is testimony to a clear resource paucity.

The second, perhaps ironically, is the growing importance of the environment as a political issue. Certainly, the present government does not see a synonymity between issues of global warming, acid rain, pollution and the like with the enjoyment of, access to and protection of the countryside. Greater attention to the former may occlude the latter. In 1989, for example, as the government was preparing its environment White Paper (Department of the Environment, 1990), the (then) Nature Conservancy Council was experiencing a cut back in its grant-aid allocation.

Third, there is an increasing marginalization of non-governmental organizations, professional bodies and pressure groups in influencing government countryside policy. The failure of government during the 1980s to produce a comprehensive White Paper on the 'countryside' or the 'rural estate' (notwithstanding some countryside issues in the recent environment White Paper (Department of the Environment, 1990)) has spawned a plethora of alternative scenarios for the countryside. Thus, the Countryside Policy Review Panel (1986), The Countryside Commission (1989b), the Council for the Protection of Rural England (1989), the Country Landowners Association (1989), the National Farmers Union (1990), the Royal Institution of Chartered Surveyors (1987), the Royal Town Planning Institute (1991) and the Town and Country Planning Association (Holiday and Green, 1991) have expressed visions, all of which move broadly in the same direction but which have not been given serious attention by government.

Historically, too, AONB and national park policy reviews have not fared well, even when produced inside government. The response to the National Parks Commission's AONB policy proposals of the early 1960s was to consider abolishing them altogether. Two out of three proposals for AONBs from Himsworth (1981) were also rejected; the third was the reaffirmation of existing policy. For national parks, even though Sandford (1974) was commissioned by government to review policy, many of its proposals were never implemented.

This does not lead to much hope for a full implementation of the two recent reports (see above) commissioned outside of government when policies eventually are distilled from them. This is particularly so when the national parks review proposes a new National Parks Act. Like the 1949 *NPAC Act* before it, this is likely to be a long time coming.

References

Bain, C., Dodd, A. and Pritchard, D. (1990) *RSPB Planscan: a Study of Development Plans in England and Wales*. Conservation Topic Paper No. 28, October, Royal Society for the Protection of Birds, London.

Benson, J.F. (1986) Integrating conservation and recreation priorities in the rural landscape. *Landscape Issues* 3, 26–40.

Blunden, J. and Curry, N.R. (1990) *A People's Charter? Forty Years of the 1949 National Parks and Access to the Countryside Act*. HMSO, London.

Box, J.D. (1991) Local nature reserves: nature conservation and public enjoyment. *The Planner* 77, 5–7.

Cherry, G. (1975) *Environmental Planning 1939–1968, Volume 2: National Parks and Recreation in the Countryside*. HMSO, London.

Cherry, G. (1985) Scenic heritage and national park lobbies and legislation in England and Wales. *Leisure Studies* 4, 127–40.

Council for Nature (1966) *The Countryside in 1970: Proceedings of the First Conference*. Royal Society of Arts, London.

Council for the Protection of Rural England (1989) *Conserving the Countryside*. The Council, London.

Country Landowners' Association (1989) *Enterprise in the Rural Environment*. The Association, London.

Countryside Commission (1977) *The Lake District Upland Management Experiment*. CCP 93. The Commission, Cheltenham.

Countryside Commission (1980) *AONBs a Policy Statement*. CCP 141. The Commission, Cheltenham.

Countryside Commission (1982) *Countryside Issues and Actions: Prospectus for the Countryside Commission*. CCP 151. The Commission, Cheltenham.

Countryside Commission (1987) *Policies for Enjoying the Countryside*. CCP 234. The Commission, Cheltenham.

Countryside Commission (1989a) *Paths, Routes and Trails: Policies and Priorities.* CCP 266. The Commission, Cheltenham.

Countryside Commission (1989b) *Planning for a Greener Countryside.* CCP 264. The Commission, Cheltenham.

Countryside Commission (1990) *National Park Supplementary Grant in England for 1991/92.* The Commission. Cheltenham.

Countryside Policy Review Panel (1986) *New Opportunities for the Countryside.* CCP 224. The Commission, Cheltenham.

Countryside Review Committee (1976) *The Countryside – Problems and Polices,* HMSO, London.

Curry, N.R. (1985) Countryside recreation priorities in the rural landscape. *Landscape Issues* 1, 4–21.

Curry, N.R. (1986) *Recreation Policy and Legislation.* Unpublished report to the Countryside Commission, Cheltenham.

Curry, N.R. (1991) *Nature Conservation Policies in Local Authorities.* Planning Gain in Nature Conservation Research Project, Working Paper 5. Department of Geography, University of Bristol.

Curry, N.R. and Comley, A. (1985) *Countryside recreation policies in structure plans.* Unpublished report to the Countryside Commission. The Commission, Cheltenham.

Curry, N.R. and Comley, A. (1986) *Who Enjoys the Countryside?* Strathclyde Papers in Planning, Department of Urban and Regional Planning, University of Strathclyde.

Curry, N.R. and Edwards, D. (1991) *Development Control in National Parks.* Occasional Paper No. 16, Centre for Rural Studies, Royal Agricultural College, Cirencester.

Department of the Environment (1967) *Parliamentary Notes on Clauses to the 1967 Countryside Bill.* Department of the Environment Parliamentary Library, London.

Department of the Environment (1990) *Our Common Inheritance.* White Paper, HMSO, London.

Dower, J. (Chairman) (1945) *Report on National Parks in England and Wales.* Cmnd 6387, HMSO, London.

Dower, M. (1978) For whom have we aimed to provide, and how is it to be achieved? In: Countryside Recreation Research Advisory Group, *1978 Conference: Countryside for All? A Review of the Use People Make of the Countryside for Recreation.* CCP 117, The Countryside Commission, Cheltenham.

Edwards, R. (Chairman) (1991) *Fit for the Future.* Report of the National Parks Review Panel. CCP334, The Countryside Commission, Cheltenham.

Fitton, M. (1979) Countryside recreation: the problems of opportunity. *Local Government Studies* July/August, 57–90.

Gosling (Chairman) (1967) *Report of the Footpaths Committee.* Ministry of Housing and Local Government, HMSO, London.

Himsworth, K. (1981) *A Review of AONBs.* CCP 140. The Countryside Commission, Cheltenham.

HMSO (1968) *The 1968 Countryside Act*, Chapter 41. HMSO, London.

Hobhouse, Sir A. (Chairman) (1947) *Report of the National Parks Committee (England and Wales).* Cmnd 6620, HMSO, London.

Holiday, J. and Green, R. (1991) *Countryside Planning*. Town and Country Planning Association, London.

House of Lords (1973) *Second Report of the Select Committee on Sport and Leisure*. HMSO, London.

Lowe, P. and Goyder, J. (1983) *Environmental Groups in Politics*. Resource Management Series No. 6, Allen and Unwin, London.

McLaughlin, B.P. and Singleton, D. (1979) Recreational use of a nature reserve: a case study in north Norfolk. *Journal of Environmental Management* 9, 213–23.

Ministry of Land and Natural Resources (1966) *Leisure in the Countryside of England and Wales*. Cmnd 2928, HMSO, London.

Ministry of Town and Country Planning (1947) *Footpaths and Access to the Countryside*. Report of the Special Committee. Cmnd 7207, HMSO, London.

National Farmers' Union (1990) *Land Use Policy Review*. The Union, London.

North York Moors National Park Committee (1984) *North York Moors National Park Plan, First Review*. The Committee, Helmsley.

Peak Park Joint Planning Board (1991) *Two Villages, Two Valleys*. Peak National Park, Bakewell.

Penning-Rowsell, E.C. (1983) County landscape conservation policies in England and Wales. *Journal of Environmental Management* 16, 211–28.

Porchester, Lord, (1977) *A Study of Exmoor*. HMSO, London.

Pye-Smith, C. and Rose, C. (1984) *Crisis in Conservation: Conflict in the British Countryside*. Penguin, London.

Roome, N. (1982) The Use of National Nature Reserves by Access Permit Holders. *Journal of Environmental Management* 14, 57–70.

Royal Institution of Chartered Surveyors (1987) *Managing the Countryside: The Policy Framework*. The Institution, London.

Royal Town Planning Institute (1991) *Rural Planning in the 1990s*. The Institute, London.

Sandford, Lord, (Chairman) (1974) *Report of the National Park Policies Review Committee*. Chairman Lord Sandford, Department of the Environment HMSO, London.

Shoard, M. (1980) *The Theft of the Countryside*. Temple Smith, London.

Siddaway, R. and O'Connor, F.B. (1978) Recreation pressures in the countryside. In Countryside Recreation Research Advisory Group, *1978 Conference: Countryside for All? A Review of the Use People Make of the Countryside for Recreation*. CCP 117, The Countryside Commission, Cheltenham.

Slee, R.W. (1982) *Country Parks: A Review of Management and Policy Issues*. Gloucestershire Papers in Local and Rural Planning 17, Gloucestershire College of Arts and Technology, Cheltenham.

Smart, G. and Anderson, M. (1990) *Planning and Management in Areas of Outstanding Natural Beauty*. CCP 295. Countryside Commission, Cheltenham.

Stephenson, T. (1989) *Forbidden Land: The Struggle for Access to Mountain and Moorland*. Manchester University Press and the Ramblers' Association, Manchester.

Woods, A. (1984) *Countryside Conservation: the Development of Policy from 1880 to 1980*. Gloucestershire Papers in Local and Rural Planning 24, Gloucestershire College of Arts and Technology, Cheltenham.

10

THE CHANGING DEMAND FOR COUNTRYSIDE RECREATION

Sue Glyptis

Research on recreation and relevant policy making in the UK has focused mainly on urban rather than rural issues. Furthermore, the literature on rural areas is dominated by urban perspectives: the rural environment is treated as a reception area, attracting recreational demand, and variously enjoying or enduring its impact. Less attention has been addressed to the rural community as a generator of recreational demand in its own right. This review encompasses both perspectives.

Countryside Recreation for Non-rural Residents

Countryside recreation is one of the most popular forms of out-of-home recreation in the UK, involving two-fifths of the population on a typical summer Sunday. Participation grows ever more diverse as new technology, the quest for challenge and adventure, and the demand for individual – as opposed to team-based – sports combine to generate new pursuits, such as jet-skiing, parascending and microlight flying. Many such activities also require purpose-built facilities, and several attract commercial interest.

Accommodating these demands would be difficult enough if the countryside were specially earmarked as a public recreation area with no competing uses. In reality, however, recreation in the UK has no such exclusive claims; it competes for land with other prime functions, notably farming, forestry, water supply, mineral extraction, military training and conservation. Recreation is a relative latecomer to the rural scene. Its impact, therefore, is not just on the physical fabric of the landscape but also on the pursuit of these other functions. The British countryside is, in no sense, purely a playground.

Nor is recreational use of the countryside a public prerogative. It constitutes use of a domain owned mainly (87%) by private individuals, and with public access dependent on certain legal rights or lenient attitudes on the part of landowners (Shoard, 1987). Even the national parks, areas designated specifically for landscape protection and public amenity (see Chapter 9), are largely in private ownership. They are not just expanses of open access land, nor are they owned or managed by the nation. A mere 17% of the total area of the parks is in public ownership. Furthermore, public owners are not necessarily guardians of amenity; 3% of the park area, for example, is owned by the Ministry of Defence. The National Park Authorities, charged with administering the parks, own a mere 0.6%.

The British countryside, then, is a remarkably private and multipurpose place; it is also far from static: the main changes in train for the 1990s are agricultural and demographic. The post-war agricultural revolution, fuelled by technological advances and bolstered by government subsidies, has produced larger farms, outputs and profits. Production, however, has outstripped demand, providing unwanted food surpluses, and as a result there are now pressures to reduce subsidies and extensify farming practices (these issues are dealt with in Volume 1 of this book). As part of this restructuring of agriculture, millions of hectares will become uneconomic to farm and available for other uses, including, perhaps, recreation. Indeed, recreation, for long perceived as a threat to the rural environment, is being seized upon as an economic saviour and an opportunity for imaginative after-use. In practice, however, limitations are likely to be imposed by farmers' understandable reluctance to be diverted from the business they know best, by a lack of capital and entrepreneurial skills to develop recreation facilities, and by residents' resistance to the building of sports halls and motocross tracks.

In many rural areas demographic changes are equally dramatic. After decades of out-migration from the countryside, the 1970s and 1980s saw a net inflow of population to rural areas, a trend likely to continue as people taking early retirement, commuters and high-technology industries are more footloose. Much of the rural scene is being progressively urbanized, and development pressures are intense. Counterurbanization brings into the countryside incomers who are affluent, mobile, articulate, and accustomed to urban standards of service provision.

Key contemporary issues for policy and research are discussed below.

Public rights of way

In the UK, public rights of way are a significant means of public access to the countryside (see Chapter 9); indeed the network of footpaths, bridleways and byways amounts to 193 000 km. The 1949 *National Parks and Access to the Countryside Act* required all county councils to identify their public rights of way and prepare definitive maps. Objections could be placed but, once entered onto the definitive map, rights of way were to run in perpetuity. As highway authorities, County Councils must maintain rights of way, whereas owners of land over which rights pass are obliged not to impede those rights by obstacles or cultivation. The paths network, though, is far from being as complete and convenient as possible.

The 1968 *Countryside Act* requires that highway authorities signpost public paths where they leave metalled roads. Shoard (1987) found that, in Gloucestershire, as many as 90% of paths lacked signposts where they left the roads, and of the 11 signposts that had been provided two pointed in the wrong direction! The Countryside Commission (1989) found that people on a 2-mile walk, planned with a map, stand a two in three chance of not being able to complete their walk because of obstacles.

Acceptable activities

Different recreational activities meet with varying degrees of welcome in the countryside. The Countryside Commission, for example, will

> focus attention on the informal aspects of enjoying the countryside. Concentrate on facilities available to the general public, recognizing that those activities organized into clubs or competitive activities are primarily the concern of the Sports Councils and the bodies represented on the CCPR; give priority to recreational activities that do not cause significant disturbances to others, recognizing that comparative tranquillity is one of the main attractions of the countryside; urge local authorities to ensure that places are available for more noisy activities.
>
> (Countryside Commission, 1987, p. 4)

Understandably, the greatest controversy is waged over national parks. MacEwen and MacEwen cite the example of the Langdale Centre in the Lake District, a prestigious time-share development with a hotel, pub, restaurant, leisure centre and access to lake and countryside:

> Is it really in the spirit of the National Park to be offering 'a holiday in the Caribbean, a tropical dream come true', where you can 'sip your drinks under the palm trees?' And what about the 'private lake frontage with four acres of land on the shores of nearby Coniston Water', in a park where the public has very limited access to the lake sides..?
>
> (MacEwen and MacEwen, 1987, p. 85–6)

The Council for National Parks believes that many forms of recreation should be forced to find a home outside the National Parks:

> Park authorities should encourage only the distinctive types of recreation which draw inspiration from the special environmental qualities of the Parks . . . the enjoyment duty should refer more

specifically to quiet, non-motorised, open air recreation. . . off-road, air-borne and water-based motorised recreational activities are inappropriate in national parks and should not be allowed.

(The Council for National Parks, 1990, p. 28)

Access, social stratification and special needs

Of all countryside trips, 68% are made by a mere 17% of the population (Countryside Commission, 1985). In absolute terms, 6–7 million people are frequent countryside users, whereas 15–16 million visit rarely or never. People having personal use of a car make up 46% of the population, but make nearly two-thirds of all countryside trips (63%). People in households without a car make up over a third (34%) of the population, but make only 18% of the trips.

Social class, too, is a powerful discriminator. People in managerial occupations are almost three times as likely to visit the countryside as those who are unemployed or on minimum incomes. Car ownership and class are, of course, interrelated, and they connect also with other factors such as income and ethnic origin.

Visitor behaviour

Much more is known of visitor characteristics than of visitor behaviour. Few surveys examine behaviour on site, and how it relates to landscape features, the siting of other facilities or the presence of other users. The few studies available show that recreational use is highly concentrated, in both time and space. Equally, therefore, impact is highly localized, at predictable times, in predictable places (Burton, 1974; Glyptis, 1981a, b). On most sites, visitors tend towards a clumped distribution pattern, resulting from a mixture of preference and site characteristics. The majority choose to stay close to the car. Many are attracted to relatively enclosed areas, such as clumps of woodland. 'Edges', such as a boundary between dune and beach or a woodland edge, also tend to attract.

Impacts of recreation

Recreation can, potentially, cause a number of adverse impacts, recently reviewed by Sidaway (1988). They include:

1. physical erosion, for instance, the erosion of hill paths through repeated use;
2. trampling of vegetation, or damage to vegetation from the wash from boats;
3. disturbance to wildlife;

4. congestion of rural roads;
5. disturbance of rural residents;
6. noise;
7. visual intrusions;
8. other social impacts – perhaps a clash of cultures between indigenous residents and incoming visitors.

Impacts, though, can be positive as well as negative. Recreational use can generate income and jobs; it can help to sustain rural service provision, road systems and transport; well designed, it can improve the appearance of the environment; and it can create new uses for derelict sites, social benefits for rural dwellers, and public awareness of environmental issues.

Countryside Recreation for Rural Residents

Insofar as rural areas feature at all in the minds of recreation policy-makers, they are generally assumed to be recreationally deprived. In traditional rural areas outmigration has left behind the less qualified, poorer, older and less mobile members of the community. The populations of many villages are inadequate to support basic community facilities. Facilities such as village halls and primary schools fall into disuse and disrepair, and the people who remain have greater distances to travel to key settlements, fewer means of doing so and, often, tortuous routes to negotiate.

However, the rural scene is changing, and recent studies in contrasting rural areas set rural recreation needs in new perspective (Glyptis 1987a, b, c). The areas examined were:

1. *Swaledale, North Yorkshire* – to the west of Richmond, mostly within the Yorkshire Dales National Park;
2. *Ryedale, North Yorkshire* – less physically remote than Swaledale, but with poor public transport provision;
3. *Castle Donington parish, Leicestershire* – between Loughborough and Nottingham;
4. *West Berkshire* – parishes around the town of Hungerford, where the population has grown steadily throughout the 1980s, and further growth is predicted until at least the mid 1990s;
5. *Theale, Berkshire* – on the outskirts of Reading, experiencing very rapid population growth.

Recreation patterns are similar in all five areas, with informal and social activities dominant. Certain local variations are evident: amateur dramatics, for example, is unusually popular in the Yorkshire areas, reflecting the strength of community endeavour based at village halls; in West

Berkshire, sports participation is higher than the national average, and in Theale it is even higher.

To examine the extent of recreational deprivation as perceived by the communities, five indicators of constraint were used. People were asked:

1. whether they had any difficulties taking part in their current leisure activities;
2. whether they had any difficulties using existing recreation facilities;
3. what they thought about the range of sport and recreation facilities currently available;
4. whether there were additional activities they would like to take part in but could not at present;
5. whether there were any additional facilities they would like to have provided.

There were few signs of massive deprivation (Table 10.1). Approximately four-fifths of people reported no difficulties doing the activities they currently took part in, and around three-quarters had no difficulty using existing facilities. Between two-thirds and four-fifths regarded existing facility provision as 'good' or 'very good'. Complaints about existing provision were greatest in Theale; so too were aspirations for future participation and provision: 37% of people wanted to take part in additional activities, and 56% wanted more facilities. By contrast, in remote Swaledale, only 28% wanted extra facilities. In this, the most rural area, residents were realistic enough to know that substantial facility provision was unlikely to be viable and, in any case, many were not keen on the idea of 'depending' on facilities, or having their leisure 'organized'.

The sense of deprivation is suppressed for several reasons. First, many people, indigenous and incomer alike, feel that their rural way of life and pleasant surroundings more than make up for the lack of facilities:

> It is a most delightful, pleasant and scenic area . . . and does not need spoiling by any more leisure opportunities.
>
> (Ryedale resident)

Sometimes there is a conflict of views between long-established residents and newcomers, mainly where the influx of newcomers has been particularly large and rapid. In some of the villages where newcomers have set up and run clubs and committees, this has been resolutely opposed by established villagers, who also complain about the 'up-marketing' of their local pubs to steak bars. Elsewhere, though, the opposite is true: established residents want to see their areas progress, and want facilities for their families, but this is strongly resisted by newcomers, who want to preserve their new-found rural paradise as they found it.

Second, local voluntary initiative plays a strong part in supplying satisfying leisure opportunities:

Table 10.1. Constraints and difficulties in participating in recreational activities.

Percentage who:	Location				
	Swaledale	Ryedale	Castle Donington	West Berkshire	Theale
Have difficulties taking part in their current activities	20	14	22	20	21
Have difficulties using existing facilities	29	17	22	25	27
Regard current facility provision as poor	25	21	31	19	32
Would like to take part in additional activities	30	24	17	29	37
Would like to have more facilities provided	28	30	27	38	56

Source: author's survey.

Our village hall is used nearly every night of the week, due to a
very good village hall committee and support from the village.

(Swaledale resident)

There is likely, though, to be more latent demand than surveys uncover,
because the very lack of provision can suppress aspirations; people make
do with what they have. To an extent supply creates its own demand:
when facilities are provided there will be a degree of usage from people
who would not have thought to ask for such a thing.

The constraints are not evenly shared. Those affected are mainly:

1. *Women* – especially those without transport, particularly young
mothers, teenage girls and elderly women. Constraints that inhibit
women's participation generally, such as lack of transport, apply all the
more acutely in rural areas because of the longer distances and greater
travel times involved in using facilities;

2. *Young people* – for children, the main deficiencies were a lack of small,
safe play areas, and a lack of play leaders. Teenagers (and their parent
chauffeurs!) had great difficulties in using town-based facilities and in
finding things to do during school holidays:

Unless youngsters have parents who are willing enough (and rich
enough) to run their children all around North West Leicestershire
there is very little for them to do.

(Castle Donington resident)

Facilities are fairly adequate, but Newsham has no kiddies'
playground or any area for older boys to play football etc. At the
moment they use the road, and the village bus shelter as a goal.
This is totally unsatisfactory.

(Ryedale resident)

3. *People without access to cars* – car ownership is unusually high in these
areas, and it is tempting to assume, therefore, that mobility is not a
problem. In reality, this is far from the case. There may be fewer people
without cars, but for the non-car owners and for the many housewives
denied the use of a car during the day there is a pervasive sense of
isolation, whether in the remote farms of Swaledale or the high-density
housing estates of Theale:

Very poor public transport. The nearest bus stop is approximately
three and a half miles away.

(Swaledale resident)

There is only one bus per week into Newbury, 10 a.m. on a
Thursday, returning at 12.30 p.m. There isn't enough time to shop
for food, let alone recreational activities.

(Theale resident);

4. *Retired people* – especially non-car owners, and those living alone.

To these four groups might be added a fifth:

5. *Newcomers* – in addition to the constraints affecting particular social groups, there can be further problems for newcomers, such as lack of awareness of opportunities, and high expectations of facility provision. There are signs that people adapt with time, either because they find ways of overcoming problems, or because they adjust their life-styles to rural circumstances. In West Berkshire, the difficulties were fewer, because many of the newcomers were older people, moving there by choice to enjoy a peaceful, rural life-style.

Adjustment is perhaps harder in the more 'urban' rural areas, where newcomers have mostly relocated because of jobs. In Theale, 42% of the newcomers had difficulties using facilities, compared with a quarter or less of those who were longer-established. Interestingly, in most cases the newcomers already had much higher than average participation rates in sport and recreation. But participation is not an indicator of contentment: they want to take part still more. Some sectors of the community, especially in the more rural areas, do little and want little. For others, abundantly represented in the new estates of Theale, the more they do, the more they want to do, and the greater their expectations that opportunities will be provided.

Whether those with the greatest aspirations can be deemed to be those in greatest need is a matter for policy-makers to decide. If those who *have* access to fewest facilities *want* fewest, there is a genuine policy dilemma: whether to spend resources providing more for the 'haves', who in their own estimation do not yet have enough, or on encouraging participation, and raising expectations, among the 'have-nots'. The former would judge themselves to be recreationally deprived; the latter would be judged so by public sector providers.

Conclusion

At the start of the 1990s countryside recreation in the UK is poised between opportunities and threats. The main opportunity is a new political and popular commitment to environmental protection, backed by a requirement that local authorities prepare detailed land-use plans, government encouragement to make positive use of vacant land, and national and local support for the creation of community forests. In addition, incentives to landowners and farmers to manage their domain in ways that protect key landscape features are meeting with encouraging response. The extensification of agriculture releases some land and buildings suitable for new recreational uses. Average levels of affluence and mobility continue to rise, providing the desire and the means for more frequent and more widespread participation in outdoor recreation.

The threats stem from the selectivity of ownership, control and access. That private ownership prevails has already been stressed; in the 1990s the private share of the total countryside resource seems set to increase further still, following the privatization of the water industry, the selling of some publicly owned forest, and the privatization of electricity. Public access may yet lose more to private profit motives. Furthermore, elitism is not confined to the landowners. It pervades many urban demand centres, as the rich get richer and the poor get poorer and less able to engage in countryside recreation. It pervades the social profile of many rural communities, as urban in-migrants create a clash of cultures with established villagers; and it sets recreational activities at odds with one another when the followers of one activity (for example, rambling) protest at the (usually legitimate) presence of others (for example, microlight flying, motocross). Planners and policy-makers need to take a comprehensive view of countryside recreation in all its many forms, and managers of individual activities need to be ready to demonstrate and disseminate methods of pursuing their recreation without detriment to the environment or other users.

References

Burton, R.C.J. (1974) *The Recreational Carrying Capacity of the Countryside.* Keele University Library Occasional Publications 11, Keele University, Keele.

Council for National Parks (1990) *A Vision for National Parks.* Evidence to the National Parks Review Panel, CNP, London

Countryside Commission (1985) *National Countryside Recreation Survey 1984.* CCP 201. The Commission, Cheltenham.

Countryside Commission (1987) *Enjoying the Countryside. Priorities for Action.* CCP 235. The Commission, Cheltenham.

Countryside Commission (1989) *Countryside Commission News* 39, September/October, The Commission, Cheltenham, p. 1.

Glyptis, S.A. (1981a) Room to relax in the countryside. *The Planner* 67, 120–122.

Glyptis, S.A. (1981b) People at play in the countryside. *Geography* 66, 277–85.

Glyptis, S.A. (1987a) *Sport and Recreation in Rural Areas. A Sample Study of Ryedale and Swaledale.* Yorkshire and Humberside Council for Sport and Recreation, Leeds.

Glyptis, S.A. (1987b) *Recreation in Expanding Residential Areas. A Study of West Berkshire and Theale.* Report to Newbury District Council and the Sports Council, Reading.

Glyptis, S.A. (1987c) *Recreation and Leisure in Castle Donington and District. Statistical Data.* Report to Castle Donington Parish Council, Castle Donington

Community College and North West Leicestershire District Council, Castle Donington.

MacEwen, A. and MacEwen, M. (1981) *National Parks: Conservation or Cosmetics?* Allen and Unwin, London.

MacEwen, A. and MacEwen, M. (1987) *Greenprints for the Countryside?* Allen and Unwin, London.

Sidaway, R.M. (1988) *Sport, Recreation and Nature Conservation.* Sports Council Study 32, Sports Council, London.

Shoard, M. (1987) *This Land is our Land.* Paladin, London.

11

TOURISM IN RURAL AREAS: CANADA AND THE UNITED KINGDOM

Richard Butler and Gordon Clark

Rural tourism has a richer history in the United Kingdom than in Canada only by virtue of the former country's longer period of settlement. Rural tourism has usually involved urban people moving to the countryside and it has progressed from an elite group favouring a few resorts to a mass phenomenon that pervades most parts of the countryside to some extent. This chapter is concerned to explore the development and key issues of rural tourism in Canada and the United Kingdom.

Some aspects of rural tourism are strikingly different in the two countries. In Canada the rural areas are seen to be synonymous with the settled agricultural districts that separate the cities from the wilderness. These Canadian rural areas have lacked until very recently a distinctive image and were not traditionally recognized as recreational areas. They were 'in between' places. In Britain, without any true wilderness, the countryside comprises all the non-urban land and has had a leisure and tourist function for a very long time. However, one very clear parallel is the confusion over concepts and definitions which complicates the studies of tourism on both sides of the Atlantic.

Concepts and Definitions

The terms leisure, recreation and tourism are sometimes incorrectly treated as synonymous. Leisure is often accepted as being a state of mind, of freedom from obligations, and is put in the context of time. Recreation can be taken as activity or inactivity engaged on a voluntary basis during leisure time for the purpose of pleasure. Tourism is a form of recreation but involves travel, and definitions often include a minimum time spent and/or distance travelled (Butler, 1989). The difference between recreation and tourism may appear a matter of degree, and indeed at times it may be difficult to separate one from the other. The activities may be identical, the only difference between a hunter who is a tourist and one who is a recreationist may be the length of time spent in the area or from how far away they have come. To a dead animal the difference may not be significant. However, to researchers the difference can be more important, involving aspects such as motivation (Pearce, 1988), behaviour (Pearce, 1982), and spending and travel patterns (Wall, 1989). The emphasis in this chapter will be on tourism rather than leisure or recreation, which reflects recent trends in both countries.

The literature on rural tourism is sparse and in both countries conceptual models and theories are lacking. Despite the unsatisfactory nature of much of the review and content of Owens' (1984) evaluation of 'Rural leisure and recreation research', his conclusion on the absence of theoretical work still stands. Many of the references in tourism are case studies with little theoretical foundation (Perdue *et al.*, 1987; Chon and Evans, 1989) or they focus on specific problems (Gartner, 1987; Gilbert, 1989). Some take a broader perspective focusing on issues and process (Chow, 1980; Middleton, 1982) for example, but very few attempt as wide an examination as Bouquet and Winter (1987). Even in such a case, theory and concepts are not the main focus, but rather implied through a collection of separate papers. There is, therefore a lack of theory and models placing rural tourism in a conceptual framework.

Rural tourism is best viewed at this time as a concept, with little deep theoretical significance. As Wall (1989) notes, tourism is often defined on the basis of the consumers, i.e. it is what tourists do, rather than on the basis of its products, as is more common for most industries. In the case of rural tourism, it is further defined in terms of where the activity occurs, namely in rural areas (however defined). Unfortunately such a definition says nothing about motivations, behaviour, effects, requirements or relationships, nor about how it compares with other elements of tourism, recreation and leisure. Some of these items will become more explicit as the phenomenon is examined in Canada and in the United Kingdom.

The Evolution of Tourism in Rural Areas in Canada

The nature of Canadian development, the absence of landed estates, and the environmental characteristics of rural areas of Canada have meant that rural tourism has remained at a low level of development for many years. The use of rural areas for tourism is summarized in Table 11.1. A major change is that only recently have the rural (as opposed to non-urban) characteristics of rural areas become of real significance to the use of these areas for tourism and recreation. Exurban tourism was located beyond the rural areas, in the wilderness or the bush. As railways opened up areas such as Algonquin Park in Ontario and Banff and other Rocky

Table 11.1. Changing aspects of the use of rural areas for tourism in Canada.

Period	Level of tourism	Developments	Activities	Impacts/Conflicts	Policies/institutional arrangements
Frontier Pre–1850	Little or none	Travellers' accommodations	Travel	None	None
Agricultural Settlement 1850–1880	Little	Travellers' accommodations domestic facilities	Bees, fairs, church-related, social	None	None
Early Cottage 1880–1900	Little, but regular	Cottages, resorts	Hunting, fishing, resort activities, pleasure travel	None apparent	None
Camping and Cottaging 1900–1920	Regular, rapid growth	Camps, parks, cottages, resorts	Hunting, fishing, cottages, travel	Some minor disturbance	Park establishment, sale of Crown land
Auto Tourism 1920–1960	Explosive growth, vast numbers	Parks, cottages, camps, resorts	Hunting, fishing, diving, water sports, golf, cottages	Competition land, water access	Park establishment, sale of Crown land
Year Round Tourism 1960–1980	Continued rapid growth	Cottage conversions, four-season resorts camps, parks	Winter sports, fishing, water golf, hiking, hunting, dude ranches	Disturbance, restrictions on activities, political control, economic	Leasing only of Crown land. Restrictions on rural severance. Pollution controls. Planning regulations
Yuppy Tourism 1980–present	Selective growth	Condominiums integrated resorts rejuvenated resorts country inns bed and breakfast farm vacations	Rural culture, festivals, crafts, winter sports, water sports, fishing, hiking, hunting, city slicking	Abandonment, emigration, lack of control	Rural conservation measures. Rural tourism promotion. Trespass/ liability action

Source: Butler (1991).

Mountain national parks in the west, the appeal of the wilderness grew and was actively marketed (Hart, 1983). Closer to the urban centres, areas such as Muskoka and the Laurentians were developed both for resorts and for private cottages.

The private cottage epitomizes the Canadian vacation from the turn of the century until well after the Second World War. 'Going to the cottage' was the Canadian equivalent of 'going to the seaside' for the British. A scaled-down version of the country residence of the aristocracy, it became a dominant feature of Canadian leisure and is found across the country. Most cottages initially were located beyond the rural areas, as were the resorts, but as more people became able to make vacations, cottage and resort developments spread to other areas, including rural locations. The key element was water; a shoreline location was and generally still is a prerequisite of the ultimate cottage site (Wolfe, 1977). Sites hitherto unsuitable for farming assumed a new value in a recreation and tourism context. As transportation evolved, the cottage changed from being a destination for the summer vacation to also being a weekend destination, and the distinction between tourism and recreation became blurred. Over the last two decades many cottages have been converted into year-round recreation facilities, and in a natural progression, often to permanent residences for commuters or retired people.

The half a million cottages in Canada in 1990 (Statistics Canada, 1990) have resulted in problems including over-use of areas, pollution, alienation of access to water bodies, traffic congestion, political unrest and local opposition. These problems have been compounded as alternative forms of residential development have taken place in traditional cottage areas and other locations (Meers, 1990). Until recently the only activities specifically on rural land were hunting and fishing, and in the last decade even these have faced opposition (Swinnerton, 1982). Snowmobilers, and to a lesser degree cross-country skiers and hikers, have also emerged as significant users of rural land, with major problems of access and trespass (Butler, 1982).

The role of the public sector in rural tourism in Canada

Origins

The involvement of the public sector in rural tourism starts in the late 19th century in Canada, with attempts to protect and prevent commercial exploitation of natural features such as Niagara Falls in the 1870s (Nelson and Butler, 1974). Such initial steps were not primarily concerned with the rural setting. The establishment of national and provincial parks similarly proceeded based on policies related to protection, maintaining representative environments, promoting regional economic development and providing leisure opportunities.

Most parks were created to provide wilderness, not rural, recreation opportunities to urban dwellers, although at least one national park (Riding Mountain) was established to provide rural recreation facilities. Eventually more parks have been provided in rural areas (for example, in the foothills in Alberta and along the Great Lakes shorelines in Ontario) but it is their location relative to cities that explains their establishment, not the attraction of the rural surroundings. The idea of specifically developing tourism and recreation in rural areas as agents of economic development was recognized at various times by the two levels of government, especially after the Second World War, when national parks were being established in eastern Canada. However, the parks were viewed as 'islands' that would generate income and jobs. They were not integrated into the rural setting.

Post-1960 involvement

In the 1960s major efforts were made to assess the capability of settled land outside the cities for a number of uses, including recreation. However, this was of little value due to an inappropriate methodology for recreation uses and to irrelevancy; no agency planned recreation and tourism at the federal level and only partly so in some of the provinces (Balmer Crapo Associates, 1975). Yet the 1970s saw a resurgence in interest by the public sector in stimulating rural tourism and recreation. While farm holidays were being promoted at the provincial level, the Canadian Council on Rural Development began studies of the implications of rural tourism and recreation for communities (Taylor, 1975). While legitimate concerns were raised over the likely impacts of large-scale urban pressures on rural areas, some of the conclusions (for example, that expected improvements in urban areas would reduce demand) were less realistic (Chou, 1975). Good critiques of the absence of clear policies relating to tourism and recreation were made by Graham (1975) and Balmer Crapo Associates (1975).

Although interest in tourism has grown at the federal level since the mid 1970s (for example, the production of a tourism policy for Canada), there has been no interest in an integrated approach to rural tourism. Provincial interest in tourism and recreation in rural areas has taken a different slant in the late 1970s and 1980s. One example of this is the review of recreational and, to a lesser extent, tourist use of agricultural land in Alberta undertaken by Swinnerton (1982), which was concerned with the pressures being exerted on the agricultural land base.

Current involvement

In the 1980s rural areas began to receive vastly increased numbers of visitors, which gave rise to two major concerns. One was the alienation of land, especially shoreline property from agriculture to non-residents. This was perceived as critical by Prince Edward Island, which amended its *Real Property Act* in 1972 to limit non-resident purchases of shore frontage (PEI Land Use Service Centre, 1978; Kienholz, 1980). In Ontario, the sale of Crown Land to non-residents and later to residents was halted, and only leases were allowed for cottage development because of similar pressures on shorelines.

The second major concern was with the abuse of rural property through trespass and problems of landowner liability (Butler and Troughton, 1985; Conservation Council of Ontario, 1975). Many rural property owners experienced problems of trespass and vandalism because of increased recreational and tourism use of rural areas, especially related to increases in accessibility caused by the introduction of off-road vehicles such as the snowmobile (Butler, 1982). In Ontario, legislation was amended significantly to reduce landowner liability and to limit access to private land (Ministry of the Attorney General, 1979, 1980).

In recent years there has been little attention paid at the federal level to tourism in rural areas. In 1990 the Federal Government published the first tourism policy for Canada (Tourism Canada, 1990) which identified Canada's 'four distinct tourism experiences' (touring, outdoor adventure, city experience and resorts) and a Federal Tourism Agenda was established for the 1990s. But there is no reference to rural Canada as an entity, nor to rural tourism. Although the policy does recognize, albeit briefly, the need to retain the integrity of environmental and cultural resources and to encourage operations along the principles of sustainable development, it argues for the promotion to the international market of what Canada currently offers.

This focus on attracting tourists to Canada is deliberate, as is the decision not to attempt to persuade Canadians to travel within Canada. It is clear that at the Federal level in Canada, the development of rural tourism is unlikely unless it can be shown to be capable of attracting significant numbers of foreign visitors. At the provincial level there is more awareness of the potential marketability of rural tourism. Farm vacations are promoted actively in Nova Scotia, Alberta and Ontario, and all the provinces promote such spring activities as blossom-viewing and maple-syrup-related activities, and, in the autumn, purchasing agricultural produce and viewing the colours. In most cases, however, the marketing and promotion of these activities are aimed at intraprovincial travellers rather than interprovincial or international visitors.

Tourism in rural areas in Canada – present and future

There is no clear concept of rural tourism nor of the role of tourism in rural Canada. The situation has changed little from what Balmer Crapo Associates described in 1975, when they noted the ignorance of impacts, a non-rural focus, a lack of framework, policy and planning, and institutional complexity. There is no sizeable body of related research in Canada as in Britain (for example, Blacksell and Gilg, 1981). In part this stems from different attitudes towards rural areas, and in part from attitudes towards tourism. Troughton noted that 'the countryside is poorly developed in Ontario in terms of its image and its functional relationships' (Troughton, 1975) and this is probably still valid for all of rural Canada. Most Canadians still view the countryside merely as something that lies between where they live and where they recreate or spend vacations. Urban residents tend to assume that no one lives there and so they can use it for garbage disposal, nuclear-power production and waste disposal, senior-citizen retirement homes and, most recently, leisure. Leisure facilities in rural areas are 'islands', and this mentality is not confined to the users. Major efforts have been made to protect and preserve the environment within parks and conservation areas, but much less effort is made outside the boundaries of these areas. The private cottage is zealously but individually guarded, and the uses associated with it often have negative effects on the surrounding area, a fact even cottage owners have begun to admit (Meers, 1990).

The absence of a clear image of the countryside in Canada has meant that tourism has begun to make major demands and cause major impacts upon parts of rural Canada. The absence of a clear understanding of tourism by many elected officials and entrepreneurs has meant it has often been encouraged and supported in ignorance of its likely effects.

As suggested in Table 11.1, there has been a shift in the appreciation of rural areas for tourism. Rural resorts have been rejuvenated and communities revitalized (Hinch and Butler, 1988) because of the rise of environmental concerns and related desires for fresh and organic products, and the trend to shorter and more frequent vacations. This trend may not be long term, indeed, given the propensity of modern tourists and recreationists to change preferences rapidly (Butler, 1989), it is highly unlikely to be a permanent shift in focus but, in the short term at least, it is likely to have increased impacts upon rural areas.

It is essential that tourism is planned and integrated into the normal pattern of activity; which is obvious and easily stated but rarely successfully achieved in the long term. Tourism is a dynamic, highly competitive and very fragmented industry that is notoriously difficult to control. The issues of capacity and limiting its development are crucial to the destination areas (Hohol, 1980). While the concept of a fixed number of visitors is unrealistic, there is little doubt that too much tourism can be more harmful and disruptive than too little.

As important as the scale of rural tourism is its nature. Some forms may be compatible with traditional activities, for instance farm vacations, but even here problems can arise between hosts and guests (Pearce, 1982, 1988), and scale is of particular significance. Other forms may be unrelated to traditional activities (for example, theme parks or marinas) but be able to exist as separate entities providing tax revenue and employment, albeit with land loss, increased visitation and service costs. The chances of conflict between tourism and other activities are high, to a large degree because of the lack of understanding of the complexity of tourism.

Control of tourism is also critical in Canada today: whether control rests with the local community may determine the success of tourism in the area. That is not to say that local control will necessarily result in appropriate tourism development, nor that local control will be permanent. However, if local control is absent or lost, the chances of development not being compatible with local preferences and needs are much greater.

Rural Tourism in the United Kingdom

Background

Tourism has had a role to play in the British countryside for several centuries. Today, however, the link between the countryside and tourism is being reinforced in Britain by the predicted decline in incomes from food production and the need to diversify the rural economy (Commission of the European Communities, 1985, 1988). Tourism is recognized as having an important role to play in the economic restructuring of the British countryside. Its distinctive features are its volume and diversity, its spatial concentration, its changing nature and the reassessment of its relative merits in economic and environmental terms.

A further problem when trying to define the scope of rural tourism in Britain is the smallness of the country compared with North America. However 'rural' is defined in the British context, the effective distances between town and country are now very small because of the motorways and high car-ownership. Much of the British countryside can be used as easily by day visitors as by those staying away from home (i.e. by tourists as conventionally defined). Similarly the growth of short breaks, which are strongly off the peak season in Britain, has reduced the formerly distinctive seasonality of rural tourism. Nonetheless, although the margins are diffuse, there is still a core of tourist activities in the British countryside which is the basis for this section of the chapter.

Rural tourism and conflict

It is paradoxical that tourism should generate so many conflicts when the nature of the activity is relaxation, enjoyment and escape from everyday tensions. In Britain the conflicts associated with rural tourism can be categorized as (i) economic, (ii) environmental and (iii) social. The nature of these conflicts is examined in turn.

Economic perspectives

The traditional British way of viewing tourism has been from an economic perspective and the verdict has normally been that tourism is a valuable component of economic development and so is worthy of official encouragement and promotion. First, tourism creates jobs in increasing numbers – 1.42 million in 1987 compared with 1 million in 1974 (British Tourist Authority, 1989). Also relevant are Archer's calculations on the not inconsiderable income multiplier from tourist expenditure in rural Britain (Archer, 1973, 1982). Whether on farms needing to diversify or in whole regions, tourism has for long been characterized as a valuable source of jobs in the face of declining opportunities in the primary sector (Bouquet and Winter, 1987; Ilbery and Stiell, 1991). A second element of the case for tourism (although not specifically for rural tourism) is its beneficial effect on the balance of payments. The UK industry's foreign-exchange earnings accounted for 4.6% of total exports in 1988 and 1.5% of GDP in 1986 (British Tourist Authority, 1989). A third element is the general buoyancy of the sector notable for the 23-fold growth in the number of foreign tourists between 1950 and 1988. Predictions suggesting continued growth, focus on the impact of the Channel Tunnel in 1993, the Single European Market, rising personal incomes and a high marginal propensity to spend money on leisure and travel. In recent years this economic case for tourism has been the subject of searching criticism. The figures for jobs in tourism are claimed to be inflated by including those involved with servicing non-tourist activities (for instance, business travel, day visitors and eating out) which have also been growing rapidly in Britain. The apparent expansion in tourist jobs will also reflect the recent contracting-out of formerly in-house functions (like catering) and new self-employed ventures induced by recession and early retirement. The barriers to getting into tourism are not so high as those elsewhere in the economy but the longevity of small businesses is usually limited.

However many jobs there are in tourism, they are not well paid – the hotel and catering industry is the second-lowest earning of any two-digit sector of the British economy (Department of Employment, 1990). Not only are hourly rates low, but seasonal and part-time working is common (particularly for women) and this further reduces their total earnings. The countryside still has a more seasonal tourism than the cities and so is even less likely to provide attractive year-round employment for local people. The need to contain costs in a highly competitive environment is cancelling out the economic benefit of increasing numbers of tourists.

Furthermore, tourism is a very unstable sector, being vulnerable to recession, exchange rates, fear of terrorism, bad weather or even a bad press or the whims of fashion. The national and international spread of hotel chains and tour companies (which can bring in extra tourists) can also lead to higher leakages of profits away from the tourist region, so reducing further the economic case.

Clearly the economic judgement on tourism in the British countryside depends on its context and form. The ideal would be a perennially popular area, run by local firms, diverse enough to minimize income leakages, and sufficiently accessible to extend its tourist season well beyond the high season. In practice the economic judgement is less favourable because of its income leakages, volatility, declining multiplier, low pay, imported labour and the conservatism of investors. The least favourable circumstance in which to promote tourism is when the rural economy is already weak, since tourism will create highly unbalanced income and employment distributions. It is better as a supplement for a thriving and diverse economy than as the mainstay of rural development.

Environmental perspectives

Whereas the economic perspective has seen a more critical view of tourism challenging the orthodoxy of its benefits, the roles are reversed when an environmental stance is taken. Tourism is seen as destructive to varying degrees, often of the very qualities that initially attracted the tourists. There are many types of environmental degradation that have been identified recently in Britain as resulting from tourism in the countryside:

1. water pollution in the Norfolk Broads from pleasure boats;
2. noise pollution from water-skiing on Windermere;
3. erosion of footpaths along the Pennine Way;
4. summer traffic congestion on roads in Cornwall.

Such problems may indeed be localized and atypical, but they are the focus for pressure groups such as the Friends of the Lake District, Tourism Concern and the Council for the Protection of Rural England. Indeed the environmental challenge has become the principal counterweight to the economic arguments.

The environmental case has been strengthened by the increasingly 'development-intensive' nature of modern tourism. Facilities are expanded to cope with more people, and standards are raised to compete with other attractions and areas. Each new facet of rural tourism, from conference centres and time-share to theme parks and war games, needs buildings and other developments. The consequent effects of an ever-expanding tourism on the landscape is a sensitive issue, particularly in the national parks. Partly this is because the change in the character of the resort leads to new and less welcome types of tourist; they are attracted to the area as much by these facilities as by the initial focus (for example, the natural beauty). The unresolved environmental issue in Britain is whether (and if so, when) the further 'spoiling' of an area should be prevented by restricting access to it (instead of enlarging the facilities). Currently restrictions are enforced only rarely (for instance in the Goyt valley in Derbyshire), although the issue has surfaced again in recent discussions on the future of tourism in the Lake District National Park.

Less attention is given in Britain to the positive environmental side to tourism. It may support otherwise uneconomic transport services. Tourism can lead to the conservation of habitats, buildings or landscapes of significance. The jobs in tourism may be poorly remunerated but they can still provide enough marginal income to stop families leaving the area. This will have major quality-of-life benefits by maintaining accessible shops and schools. Some people may migrate into the area because of their enjoyment of it through holidays, and they may set up businesses and create employment.

In this environmental debate there are two opposed viewpoints. Some people seek a personal experience of quietness and an experience of nature, while denying these to more than a few others. They reject the development-intensive nature of modern tourism. The other group believes in improving on the 'natural' and in holidays having standards of living at least as good as those at home. The British environmental debate is as much about these social and cultural interpretations of the environment as it is directly about the physical environment.

Social perspectives

If the economic and environmental perspectives provide ample material for debate, the social perspective gets closer to what is commonly understood by conflict. The UK is a small, densely populated country by the standards of North America, most of it privately owned and almost none of it wilderness in any true environmental sense. The potential for social conflict is much greater in Britain, as is witnessed by some examples:

1. mass trespasses of ramblers across private moorland in the Pennines;
2. local opposition to holiday cottages and second homes in parts of Wales;
3. the enduring distinction between respectable tourists and day-trippers (see Walton, 1983).

Holidays have always reflected and reinforced the social divisions in

British society. Holidays also mark out social groups in a way that cuts across conventional work-based social classifications (Warde and Savage, 1991). Divisions can also appear between the host communities and the tourists, and between factions within the host community. Some people welcome tourists as a source of jobs and income, whereas others reject or fear them as strangers from the cities who are culturally different or threatening. Resistance to tourists can also come from recent in-migrants to the area; they have been attracted by the same qualities that are drawing the tourists (Champion, 1989). The stage across which these social conflicts are played out is usually that provided by the planning system.

Planning and rural tourism

In the middle of this complex web of social conflicts sits the British planning system. Local planning committees are entrusted with the task of granting planning permission for proposed developments. In the tourism field, as elsewhere, they have had to resolve the central dilemma of planning in the 1980s. 'Thatcherism' leans towards allowing tourist developments in the name of supporting free enterprise, reducing the countryside's need for public subsidy, and helping small firms (Banks, 1987). However, the rise of environmentalism up the political agenda has favoured restricting some tourist developments, despite the hierarchy of local and national tourist boards which exist to promote it (National Parks Review Panel, 1991). The solution to the dilemma has been complex: in areas such as national parks and green belts the conservationist argument has tended to win; in other areas development has gone ahead. Implicit in this solution is a spatial hierarchy of countryside based on scarcely acknowledged aesthetic judgements on landscape quality. Working against the restrictive stance should have been the knowledge that restricting the supply of development land can be very inflationary (Hall *et al.*, 1973; Clark, 1982). But this view has been only weakly effective in Britain: too many groups have gained from the way planning minimizes risks and enhances the capital value of land and houses. The tactic of spatial zoning has been one of the British planners' best tools for resolving conflicts. It can be seen at work keeping apart incompatible activities on Windermere, in the Cairngorms and on the Norfolk Broads. The problem of how to zone (how much space for each and where it should be) is decided pragmatically rather than by the use of strict rules. The difficulty comes when one group wants to remove another completely – second homes from Snowdonia or water-skiers from Windermere, for example.

The other area where British planners have had some room for manoeuvre has been in whether to concentrate tourism into 'honey pots' or let it spread widely. The balance of advantage seems to lie initially with concentration to get to a critical mass (for example, the

Fig. 11.1. Tourism priority areas in Cumbria. Source: Cumbria Tourist Board (1987). Reprinted with permission from Cumbria Tourist Board

Aviemore skiing resort), and then to disperse new developments when over-concentration leads to environmental degradation, as in parts of the Lake District (Fig. 11.1).

Conclusions

Despite the quite distinctive paths followed by rural tourism in the two countries, it is clear that there have been some powerful themes common to them both. One such is the persistent unevenness of tourist development. Although a veneer of holiday-making can be found throughout the countryside, the growth of tourism since 1945 has reinforced the well-established dominance of traditional areas such as the Rockies and the

Scottish Highlands. There have been some casualties; foreign holidays have reduced considerably the popularity of the British seaside resort. Yet the bulk of new investment has reinforced the 19th century pattern of holiday areas in Britain and has not seriously challenged the pre-eminence of the Federal National Parks and the areas of 'wilderness with cottages' in Canada.

A second theme is the increasingly national and international scale of the industry. Although this trend has been more fully developed in Britain, ever-widening competition and the non-local sources of investment capital are inducing a greater concern for image-making, advertising and other methods of 'product differentiation' among locations for rural tourism. The aim is to separate the countryside from the mass-market and tight profit margins of the Mediterranean and the Caribbean.

Fundamental to the success of rural tourism is an appreciation of the nature of tourism and the varying viewpoints of the actors involved (Mormont, 1987). If tourism in rural areas is to achieve a long-term relationship with traditional elements in rural areas, more careful study will be needed before development occurs, and procedures for integrating and controlling tourism should be put into place. Encouraging tourism development in rural areas, simply because it is economically feasible, should not be acceptable if tourism is meant to do more than provide economic returns for mostly non-rural entrepreneurs and leisure opportunities for the urban populations.

A criticism of tourism studies in both countries is their lack of social concern. Tourism geography is the study of those who take holidays: those who do not are rarely considered. Only 8% of British managerial and professional people did not take a holiday in 1988, compared with 45% of the unskilled and pensioners (up six percentage points since 1970) (BTA, 1989). The reasons for not taking a holiday may include work commitments, illness, disability, poverty or family ties. Tourism reflects social divisions of wealth, health, gender and family roles, both in terms of who takes holidays and their type of holiday. The role of tourism as an expression of social structure has not been well handled by geographers, whereas social historians have been much more aware of the issue (Walton, 1983).

Another theme which has not been dealt with properly is the commodification of tourism. A holiday is a curiously mixed product to buy, since it comprises not only specific items and services, such as food and travel, but also getting others to do what you would usually do for yourself (for instance cooking). Fierce price competition means that a holiday package now also includes doing things for yourself again (for example making your own coffee). Yet tourism has always been recognized as more than just a package of real goods and services. It includes relaxation, novelty, education and excitement, which can all be converted into real priced commodities so partially negating the price competition of the basic package. Relaxation becomes buying a sauna rather than sitting on a deck chair; education needs an entrance fee to a museum or heritage centre rather than walking round the area yourself.

Since tourism has always been about buying dreams, another feature of the industry is the increasing importance of advertising to differentiate packages, services and places at local, national and international scales (Urry, 1990). Such market fragmentation allows places to stand out in the increasingly crowded market place as new locations take to tourism, for example, Wigan for industrial archaeology, Glasgow for cultural tourism, and more distant places come within reach (for instance, the Far East). The major tour companies are competing on price for the mass market and levelling down to standard types of package which can be delivered with maximum efficiency. Yet tourism is partly a fashion item that resists uniformity of style on the demand side. Congestion of existing resorts, and the desire to gain status by being different, will eventually lead even the most gregarious to seek out something new. On the supply side, differentiation of products is a reaction to these levelling tendencies as small companies seek to create new holiday products for which they can be price setters.

A final area where some research is needed is in the changing relationship between tourism and its host community. Rarely is tourism the sole rural economic activity. Over the last few decades the countryside has witnessed major changes in its social composition, the main symptoms being gentrification, new forms of social polarization, and a domination by the service class. More research is needed on the relationship between the uneven social recomposition of the countryside, the spatially variable development of tourism, and the problematic relationship between the two.

This chapter has set out some of the changes in rural tourism viewed from economic and environmental perspectives. In both cases cogent arguments are now being raised against placing too much reliance on modern forms of mass tourism, since these can lead to development which is unbalanced in terms of gender, space and income distribution, and which can have such major effects on the environment. Additionally, the potential of the study of rural tourism to illuminate the development of theoretical aspects of human geography seems to have been under-valued. Rural tourism offers a rich and under-utilized laboratory for the study of restructuring, post-Fordism and the neglected cultural dimension in geographical writing, particularly in the area of personal and group images of rurality.

References

Archer, B.H. (1973) *The Uses and Abuses of Multipliers*. Tourism Research Paper 1, Economics Research Unit, Bangor.

Archer, B.H. (1982) The value of multipliers and their policy implications. *Tourism Management* 3, 236–41.

Balmer Crapo Associates (1975) *Federal Programs Affecting the Development of Recreation and Tourism in Rural Areas*. Canadian Council on Rural Development, Ottawa.

Banks, R. (1987) *New Jobs from Pleasure*. Conservative Party, London.

Blacksell, M. and Gilg, A. (1981) *The Countryside: Planning and Change*. George Allen and Unwin, London.

Bouquet, M. and Winter, M. (eds) (1987) *Who From Their Labours Rest: Conflict and Practice in Rural Tourism*. Gower, Aldershot.

British Tourist Authority (BTA) (1989) *Digest of Tourist Statistics – 13*. British Tourist Authority, London.

Butler, R.W. (1982) The development of snowmobiles in Canada. In: Marsh, J. and Wall, G. (eds) *Recreational Land Use: Perspectives on its Evolution in Canada*. Carleton University, Ottawa, pp. 365–90.

Butler, R.W. (1984) The impact of informal recreation on rural Canada. In: Bunce, M.F. and Troughton, M.J. (eds) *The Pressure of Change in Rural Canada*. York University Geographical Monograph 14, Downsview, Ontario, pp. 216–41.

Butler, R.W. (1989) Tourism and tourism research. In: Jackson, E.C. and Burton, T.L. (eds). *Understanding Leisure and Recreation*. Venture Publishing, State College, Pennsylvania, pp. 567–96.

Butler, R.W. (1991) *The City in the Country: Policies for Compatible Tourism*. Paper presented at the Annual Meeting, Institute of Public Administration of Canada, Halifax, Nova Scotia.

Butler, R.W. and Troughton, M.J. (1985) *Public Use of Private Land*. University of Western Ontario, London.

Champion, A.G. (1989) *Counterurbanization*. Edward Arnold, London.

Chon, K.-S. and Evans, M.R. (1989) Tourism in a rural area – a coal-mining-county experience. *Tourism Management* 10, 315–21.

Chou P.H.N. (1975) *Trend of Urban Demand for Tourism and Recreation in Rural Areas*. Canadian Council for Rural Development, Ottawa.

Chow, W.T. (1980) Integrating Tourism with Rural Development. *Annals of Tourism Research* 7, 584–607.

Clark, G. (1982) Housing policy in the Lake District. *Transactions of the Institute of British Geographers* 7, 59–70.

Commission of the European Communities (1985) *Tourism in the European Community*. European File 11/85, Brussels.

Commission of the European Communities (1988) *The Future of Rural Society*. COM(88) 501, Brussels.

Conservation Council of Ontario (1975) *Private Land, Public Recreation and the Law*. The Council, Toronto.

Cumbria Tourist Board (1987) *A Vision for Cumbria: Tourism Strategy*. Cumbria Tourist Board, Windermere.

Department of Employment (1990) *New Earnings Survey 1990, Vol. A*. HMSO, London.

Gartner, W.C. (1987) Environmental impacts of recreational home development. *Annals of Tourism Research* 14, 38–57.

Gilbert, D. (1989) Rural tourism and marketing. *Tourism Management* 10, 39–50.

Graham, W.W. (1975) *Provincial Programs Affecting the Development of Recreation and Tourism in Rural Areas*. Laurentian Institute, Ottawa.

Hall, P., Thomas, R., Gracey, H. and Drewett, R. (1973) *The Containment of Urban England*. Allen and Unwin, London.

Hart, E.J. (1983) *The Selling of Canada*. Altitude Press, Banff.

Hinch, T. and Butler, R.W. (1988) The rejuvenation of a tourism centre: Port Stanley, Ontario, *Ontario Geography* 32, 29–52.

Hohol, F. (1980) Communities in transition: Elmira and St. Jacobs, Ontario. Unpublished Master's thesis, Wilfrid Laurier University, Waterloo.

Ilbery, B. and Stiell, B. (1991) Uptake of the Farm Diversification Grant Scheme in England. *Geography* 76, 259–63.

Kienholz, M. (1980) *The Land-Use Impacts of Recent Legislation in PEI*. Lands Directorate, Ottawa.

Meers, D.E. (1990) The use of a commercial tourism/recreation resource. Unpublished Master's Thesis, University of Western Ontario, London.

Middleton, V.T.C. (1982) Tourism in rural areas. *Tourism Management* 3, 52–8.

Ministry of the Attorney General (1979) *Discussion Paper on Occupiers' Liability and Trespass to Property*. Ministry of the Attorney General, Toronto.

Ministry of the Attorney General (1980) *Property Protection and Outdoor Opportunities*. Ministry of the Attorney General, Toronto.

Mormont, M. (1987) Tourism and rural change: the symbolic impact. In: Bouquet, M. and Winter, M. (eds) *Who From Their Labours Rest: Conflict and Practice in Rural Tourism*. Gower, Aldershot, pp. 35–44.

National Parks Review Panel (1991) *Fit for the Future*. Countryside Commission, Cheltenham.

Nelson, J.G. and Butler, R.W. (1974) Recreation and environment. In: Manners, I. and Mikesell, M. (eds) *Geographical Perspectives on Environment*. Association of American Geographers, Washington DC, pp. 190–210.

Owens, P.L. (1984) Rural leisure and recreation research: retrospective evaluation. *Progress in Human Geography* 8, 157–88.

Pearce, P.L. (1982) *The Social Psychology of Tourist Behaviour*. Pergamon Press, Oxford.

Pearce, P.L. (1988) *The Ulysses Factor*. Springer Verlag, New York.

Perdue, R.R., Long, P.T. and Allen, L. (1987) Rural resident tourism perceptions and attitudes. *Annals of Tourism Research* 14, 420–9.

Prince Edward Island Land Use Service Centre (1978) *Non Resident Land Ownership Legislation and Adminstration in Prince Edward Island*. Lands Directorate, Ottawa.

Statistics Canada (1990) *Household Facilities and Equipment*. Ministry of Supply and Services, Ottawa.

Swinnerton, G.S. (1982) *Recreation on Agricultural Land in Alberta*. Environment Council of Alberta, Edmonton.

Taylor, G.D. (1975) *Social Implications of Recreation and Tourism for Rural People*. Canadian Council for Rural Development, Ottawa.

Tourism Canada (1990) *Tourism on the Threshold*. Ministry of Supply and Services, Ottawa.

Troughton, M.J. (1975) Agriculture and the countryside. In: Troughton, M.J., Nelson, J.G. and Brown, S. (eds) *The Countryside in Ontario*. University of Western Ontario, London, pp. 45–77.

Urry, J. (1990) *The Tourist Gaze: Leisure and Travel in Contemporary Society*. Sage, London.

Wall, G. (1989) *Outdoor Recreation in Canada*. Wiley, New York.

Walton, J. (1983) *The English Seaside Resort: a Social History 1750–1914*. Leicester University Press, Leicester.

Warde, A. and Savage, M. (1991) *The New Urban Sociology*. Macmillan, Basingstoke.

Wolfe, R.I. (1977) Second homes purpose built for an inessential purpose. In: Coppock, J.T. (ed.) *Second Homes – Curse or Blessing?* Pergamon, London, pp. 184–211.

IV

THE CHANGING EMPLOYMENT BASE OF RURAL AREAS

IV

12

AGRICULTURE IN THE CONTEXT OF THE RESTRUCTURING OF RURAL EMPLOYMENT

Nigel Walford

A crucial element in the development of capitalist agriculture in the post-Second World War era has been rising productivity based on increasing mechanization and a reducing labour force. Other factors have contributed to the trend, including governmental support for the industry, improvements in yields through scientific advances and greater stress on farm management. They have all played their part in bringing about what is now seen by some writers (Britton, 1968) as a remarkable growth in productivity, and by others as a crisis in agriculture (Munton *et al.*, 1990). This crisis has a number of components including structural production surpluses, declining real income from farming, consumer demand for better food quality, a reaction against environmentally damaging agricultural practices and public concern for conservation of the countryside (these issues are dealt with in Volume 1 of this book).

The post-Second World War consequences of agricultural change on employment, and on demographic and social structure, were extensively researched (Saville, 1957; Lawton, 1968). The predominant direction of population movement was from the countryside to the burgeoning cities and towns; the population of the countryside became more sparse and, in extreme cases, farms were abandoned with some settlements dwindling to a residual level. During the last two decades this trend has in general been reversed, with the emergence of a phenomenon variously known as 'population turnaround', 'counterurbanization' or 'rural repopulation' (Champion, 1981, 1982; Fielding, 1982; Grafton, 1982). In spite of the enthusiasm for detecting this trend, population levels in some rural areas are still declining or at best remaining stable. Furthermore, a turnaround in the direction of population movement does not necessarily invigorate specific communities, let alone allow them to return to their former economic composition. Very often the rural areas with the greatest population

growth rates are those lying within the commuting and functional hinterland of the metropolitan and urban centres.

Counterurbanization has contributed to the development of a new
functional relationship between town and countryside, which is impacting
on rural labour markets. In addition, agricultural employment has been
subjected to further reorganization. For example, whole-time employed
labour has declined, whereas the more ephemeral part-time, seasonal and
family labour has shown some increase (Walford, 1981; Ball, 1986). In
other words there has been substitution of permanent by temporary or
short-term labour. Furthermore, the farm worker has become increasingly
specialized: the traditional general-purpose farm labourer has been largely
relegated to history and popular mythology. Modern agriculture not only
demands sophisticated machinery, but also a skilled workforce to operate
this machinery. More recent examinations of labour on farms have tended
to focus on new questions, for example gender divisions both in the farm
workforce and the farm household (Gasson, 1984; Little, 1991; Symes,
1991).

The debate about the form of the countryside and rural economy that
will emerge during the 1990s is not yet concluded. For example, proposals
for more environmentally sensitive farming are instinctively countered
with a reassertion of traditional views on productivity and efficiency from
the old guard reactionaries. Even if the outcome is as yet unclear, there
seems little doubt that a far-reaching restructuring of rural areas is taking
place (Munton *et al.*, 1990; Walford, 1991). This chapter has two main
aims with regard to this process of change. The first is to examine the
general context of the restructuring of rural employment in Britain; the
second is to consider the position of the agricultural labour force within
that context under the pressures currently facing the industry. However,
more important than an examination of the empirical evidence is the
debate about whether policies intended to achieve agricultural objectives
might also affect rural employment. Policies promoting diversification or
maintenance of farming in marginal areas might help to arrest the decline
in employment, but the fate of the farm worker under the more extensive
and diversified systems of farming that have been mooted, and that are
indeed being promoted by such measures as Environmentally Sensitive
Areas, is at present largely a speculative issue.

Agriculture and the Context of Rural Employment Change

Conceptual questions

It is important to clarify terms and to establish what is meant by 'rural'
employment and the 'restructuring' of such employment. The first term

can either refer to employment in rural areas or to those forms or categories of employment that are rural in character. In other words, is there something particular and distinctive about labour markets in rural areas? Do the processes operating in these areas, and the patterns and changes in employment which they produce, differ significantly from those occurring in urban or metropolitan areas? Alternatively, are there some types of employment which are specifically rural, not because of some locational or geographical attribute, but because of the nature of the economy to which they belong? The problem with trying to identify rural employment in this way is that agriculture, forestry and extractive industries are selected because they are what one expects to find in a rural area. If this is the case, then the distinction between the two types of definition has collapsed, because the importance of place has emerged. This discussion inevitably takes us back to the well-known debate about applying the descriptor 'rural', and the concept of 'rurality', to places or to more abstract phenomena such as employment.

For example, Newby (1977), in his examination of *The Deferential Worker*, embarked upon the exposure of the myth that the agricultural workforce was generically different from that in other industries. True enough, trade unionism was relatively ineffective in the industry. On other hand, it was argued that relationships between employer and employee were becoming more like those in other industrial sectors. Indeed any differences were growing less apparent as farming evolved into agribusiness.

Unfortunately, the definitional debate does not end with the term 'rural'. It is increasingly recognized that economic and social systems are not geographically specific, but are national and international in character. The place of rural and urban space as an explanatory variable has been reduced, although there is a continuing need to use spatial divisions as a framework for understanding the processes operating in an area that would otherwise be geographically undifferentiated (Cloke, 1987; Hoggart, 1990). Such frameworks comprise a starting point for description rather than a foundation on which explanation can be grounded. Hodge and Monk (1991), for example, recently went *In Search of a Rural Economy* and concluded that 'the pattern of economic activity in many areas which are "physically" rural is intimately bound up with the wider pattern of economic activity' (p. 53). They suggest that such areas 'are all part of some larger economic system' (p. 53), which is nevertheless by implication subject to spatial partitioning.

There have been many attempts to define and delimit rural areas (see discussion in Walford and Hockey, 1991). Some relatively simple, but nonetheless effective methods, use a single criterion variable, such as population density, with a considered but nonetheless arbitrary cut-off point between urban and rural. Other more sophisticated statistical procedures, such as cluster and principal component analyses, interrelate a selection of social, demographic, economic and sometimes geographical variables in the computational melting pot. The approach is essentially multidimensional, so that the geographical units being classified are allocated among a number of categories. The identification of 'coastal retirement', 'southeastern growth' and 'parks and countryside' areas, among others, in Hodge and Monk (1991), exemplifies the methodology.

A second methodological issue relates to the nature of 'restructuring' in rural employment change. At one level, so far as employment is concerned, the concept refers to the changing fortunes of different industries, which results in relative adjustments in their levels of employment. However, there are other issues to be considered. The term 'restructuring' seems to imply that the framework within which an activity occurs has altered. With regard to employment, this might relate to a requirement for different types of skills in a workforce and a reorganization of how the business is capitalized, with greater national and international connections. Another issue is the spatial impact of such restructuring – the concern of this chapter – which looks at the structure of rural employment and the changes in the agricultural workforce.

Rural employment change

The last two decades have been a turbulent time so far as employment in Britain is concerned. In the period 1971–81 employment declined in the primary, manufacturing and construction sectors, whereas there was strong growth in service sector industries (Champion *et al.*, 1987). In 1971, 41.6% of employment in Great Britain was in the service industries; by 1981 this had risen to 44.6% and by 1989 to 48.6%. But even this sector is not immune to the effects of recession on levels of employment: in the 1978–81 period, the level of growth was only 1% compared with 12.3% between 1971 and 1978 (Champion *et al.*, 1987).

During the boom years of the 1980s there was again a growth in certain sectors of employment (Fig. 12.1), although Champion *et al.* (1987) noted that growth in service sector jobs compensated 'for only a small part of the employment shed by the manufacturing sector'. Overall, employment declined between 1981 and 1989, although the services and distribution sectors showed even higher gains than in the preceding decade. Early in the most recent period of recession, June 1989 to December 1990, official statistical sources suggest that the overall decline in employment had been fairly modest (less than 0.1%), although more recent information suggests that the fall may have accelerated during 1991. The pattern of change for the different sectors in the June 1989 to December 1990 period is not entirely consistent with previous trends, since employment in the metals, minerals and chemicals, and the distribution, hotels, catering and repairs

Table 12.1. Agricultural labour in England and Wales, 1960–89.

	Regular hired and family workers		Farmers, partners and directors		Seasonal casual workers		Total
	Whole-time	Part-time	Whole-time	Part-time	Male	Female	
Change % 1960–69	−38.4	−19.2	NA	NA	−34.7	−20.0	NA
Change % 1970–79	−28.5	−17.4	+ 5.0	+45.2	+27.8	+ 5.7	− 4.9
Change % 1980–89	−26.6	− 9.4	−11.7	+11.6	− 6.1	−24.4	−10.7

NA: not available.
Source: Derived from MAFF statistics.

industrial divisions experienced increases of 6.3 and 5.2% respectively. Apart from construction, where there was a slight increase, employment in all other sectors decreased between 1989 and 1990. In particular, agriculture, forestry and fishing continued their downward spiral, falling by 5.3%. Over the two decades (1971 to 1990) the different patterns of growth and decline in the various industrial sectors have altered the structure of employment in Britain.

During the 1971–81 period, the two most stark features of regional differentiation in employment change were: pronounced falls in areas of population concentration, and the outward spread and growth of jobs in southeast England and rural areas in Scotland, Wales and southwest England. Champion *et al.* (1987) recorded that the former areas had a decline in excess of −8.4% over the decade, while the latter had gains of over 13.4%. These authors also showed that the 'highest rates of growth [in service sector employment were] mostly in the smaller cities and towns, and some rural areas' (p. 70) between 1971 and 1981. Employment in producer services (insurance, banking, accountancy, advertising and other 'professional' services) in Britain increased by 35.6%, while in the city, town and rural local labour markets the corresponding increases were 60.2, 57.9 and 52.0% respectively.

Champion *et al.*'s (1987) study was based on functional areas which, although published at the start of the 1980s, were defined by applying algorithms to the 1971 Population Census statistics, linked with other data for the 'old', pre-1974/75, local authority areas. New functional regions based on the 1981 Census statistics were not produced. The definition of rural and urban areas for the 1980s, used in Fig. 12.2, demonstrates the application of two contrasting methods of classification to local government districts in England and Wales. The two methods are: Cloke's reworked rurality index (Cloke and Edwards, 1986), and the population density measure used by the Countryside Change Programme (Walford and Hockey, 1991). Different statistical sources were used to prepare the information presented in Fig. 12.2, namely the 1981 Population Census and the 1987 Census of Employment. These surveys operate on different bases and there is some incompatibility in their data collection procedures. However, even if the magnitude of the changes is not precisely accurate, it is likely that the direction of change is a reasonable indication of what has happened to rural employment between 1981 and 1987.

The two methods of rural/urban classification give remarkably consistent results for employment in the agriculture, forestry and fishing, and the energy and water supply industrial divisions in both years. Approximately 5.0% and 2.5% of employees in the rural areas of England and Wales worked in these two divisions in 1987. These figures are slightly lower than the 1981 equivalents (7.0 and 3.0% respectively). In other industrial divisions, which have necessarily been grouped because of data

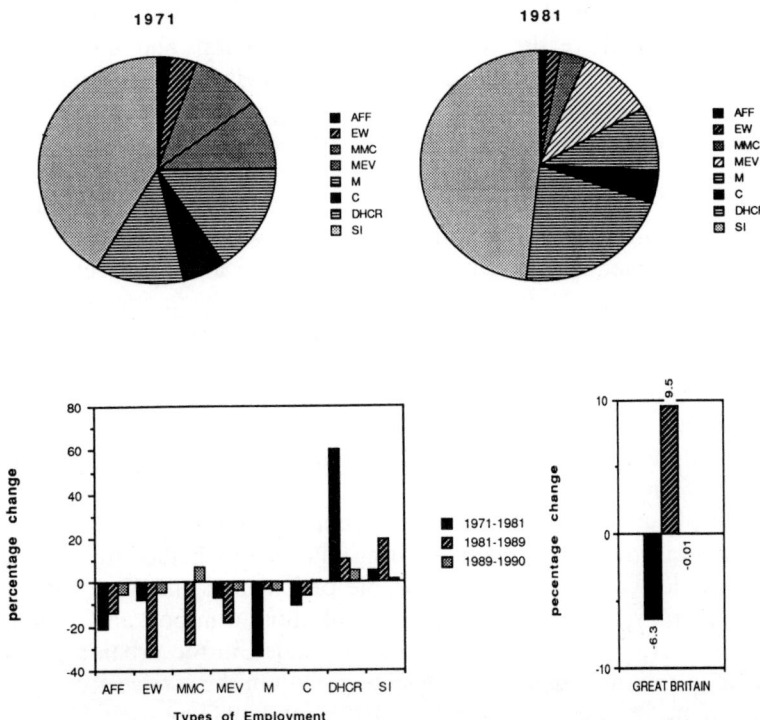

Fig. 12.1. Employment change in Great Britain. AFF – Agriculture, Forestry and Fishing (0); EW – Energy and Water Supply (1); MMC – Metals, Minerals and Chemicals (2); MEV – Metal Goods, Engineering and Vehicles (3); M – Other Manufacturing (4); C – Construction (5); DHCR – Distribution, Hotels, Catering and Repairs (6); SI – Service Industries (7–9). Note 1: The 1971 figures for industrial divisions 2 and 3 are combined. Note 2: Numbers in ( ) refer to standard industrial division codes. Note 3: Service Industries includes Transport and Communication, Banking, Finance and Business Services, Public Administration and other services. Source: Population Census and Census of Employment Statistics in Regional Trends (1990) and Employment Gazette (July 1990).

availability, there are differences between the two systems of classification, more especially for the 1987 figures. The differences are relatively small – only in the region of 4 percentage points – apart from the case of the service sector, where the combined 'Extreme Rural' and 'Intermediate Rural' districts in the rurality index was 46.8% in 1987, compared with 38.1% for the rural area defined by means of population density.

The change in the structure of employment in rural and urban areas between 1981 and 1987 is also shown in Fig. 12.2, although differences between the statistical sources and changes to the Standard Industrial Classification mean that the results should be treated with caution. Service sector employment increased by 16% whereas in all other sectors there was decline, most notably in energy and water supply, and manufacturing (−29.6% and −16.2% respectively). In general the rural and urban area aggregates, according to both systems of classification, have followed the changes recorded for England and Wales as a whole. The growth of service sector employment is apparent for the rural and urban areas, although there are differences of magnitude between the two classification systems.

The information presented in Figs 12.1 and 12.2 present two slightly different pictures of employment restructuring. Rural areas have shared in the overall decline in manufacturing, although whether this was to a greater or lesser extent than the whole of England and Wales is unresolved. Employment in construction, distribution, transport and communication industries (divisions 5 to 7) has expanded in the urban areas, but declined in the rural aggregate. Despite growth in the service sector, rural employment declined overall between 1981 and 1987 by some 8%, while in the urban area the increase was just under 2%.

Restructuring the Agricultural Workforce

The agricultural workforce has been reducing for many years: in 1881 agricultural workers comprised 10% of employment, whereas they now account for less than 2%. Even in some of the agricultural heartland areas, such as parts of East Anglia and the East Midlands, the figure rarely exceeds 10%. There are some distinctive features of agricultural employment: family businesses dominate the industry, a high proportion of the owners of farm capital do manual work and there is a larger number of individual production units than many other industries. Furthermore, the typical division of labour into owners, managers and workers is more difficult to apply, since individuals may have overlapping functions, and in many cases members of the farm family share the owning, managing and labouring functions. The industry also has a growing number of large agribusinesses where the divisions of labour found in other industrial enterprises (directors, managers, office staff, foremen and workers) are

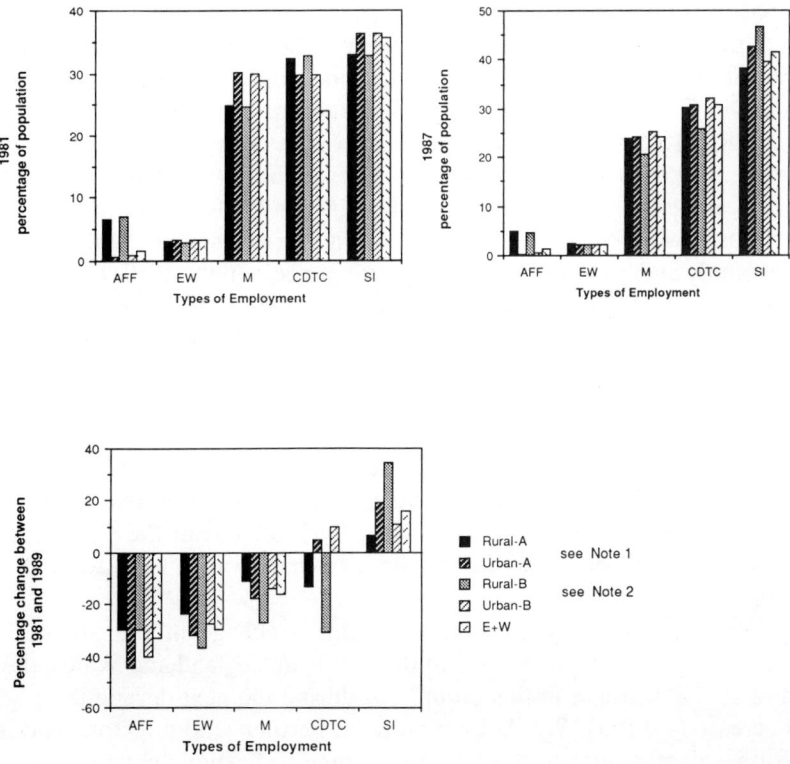

Fig. 12.2. Employment change in rural and urban areas of England and Wales. AFF – Agriculture, Forestry and Fishing (0); EW – Energy and Water Supply (1); M – Manufacturing (2–4); CDTC – Construction, Distribution, Trans. and Communication (5–7); SI – Service Industries (8–9); E+W – England and Wales. Note 1: Countryside Change Population Density Measure (4 persons per hectare). Note 2: Cloke's Rurality Index 'Extreme Rural' and 'Intermediate Rural' combined. Note 3: Numbers in ( ) refer to standard industrial codes. Note 4: Service Industries includes Transport and Communication, Banking, Finance and Business Services, Public Administration and other services. Source: Population Census and Census of Employment Statistics in Regional Trends (1990).

present. These features make it difficult to derive an accurate measure of the human input into agriculture, let alone to differentiate between management and manual labour.

Difficulties of measuring the downward trend in the size of the agricultural labour force cannot disguise the decline. Indices of change for various groups in the workforce have been calculated for England and Wales from Agricultural Census statistics (Table 12.1). A discontinuity in the time series in 1969/70 leads to problems in examining the whole 1960–1989 period, nevertheless the rate of decline in the 1970s more than doubled in the following decade. The reduction in the number of regular whole-time family and hired workers is very pronounced: in the 1960s the number decreased by 38.4%, in the 1970s by 28.5% and in the 1980s by 26.6 %. Part-time regular workers have shown slightly more resilience and, although they have declined, the annual rate of loss has been lower, falling to under 1%. Phillips and Williams (1984) record whole-time family employees as having 'fallen by half during the 1970s', and part-time family workers almost as quickly. Agriculture is traditionally a male-dominated industry and the proportion of females in the regular full-time workforce in England and Wales fell from 8.4% in 1955 to 7.6% in 1969; in the new series there has been a similar decline, from 10.6% in 1970 to 9.5% in 1980.

Table 12.1 shows that there has been a change in the number of farmers, partners and directors. In the 1970s, in England and Wales, there was a 13.4% increase in this group, but during the next decade there was a decrease (−4.8%). Whole-time farmers, partners and directors working full-time decreased, although there is some indication that the number involved on a part-time basis with their farm has been increasing. So there is a fairly clear indication of a contraction in the number of farmers. It is probably too early to say whether the increase in part-time farmers is only a temporary development, but the interest in farm diversification and alternative sources of income makes the trend very significant.

Fluctuations in the number of seasonal and casual workers (Table 12.1) have been an important change in the agricultural workforce. This is a heterogeneous group that at one time included gangs of city dwellers who migrated seasonally to work on farms, but is now composed mainly of gypsies, students, spouses of regular workers and other local residents. Many farms employ casual workers in small numbers, and the increase in this element in the total labour force during the 1970s suggests that farmers have been using this type of labour in place of having a large number of, or any, regular workers. The trend towards using seasonal and casual labour in place of regular labour does not appear to have continued in recent years: the number of male and female seasonal and casual workers in England fell by 6.1 and 24.4% respectively during the 1980s. It is also significant that the number of women employed in this way on farms fell by nearly a quarter.

It will inevitably take time for some recent socioeconomic pressures and influences to have an effect on employment in agriculture, and still longer for the implications to be revealed in official statistical sources. Farmers as individuals react in different ways, but it is difficult to see how the current pressures can stimulate much new employment in the industry. Indeed Errington (1988) has suggested that a further reduction of the agricultural workforce can be expected to help to alleviate existing 'disguised unemployment' in the industry. Nevertheless, as Table 12.1 shows, some effects can already be detected. In particular, the post-Second World War decline in the agricultural labour force, which was connected with increasing productivity, has been given a further impetus during the 1980s.

Future Prospects for Agricultural Employment

An analysis of past trends in agricultural employment is undertaken not in order to predict the future but to show the contradictory fluctuations that can occur over time. This is a result, partly, of the arbitrary or pragmatic selection of time periods, and partly of the innate difficulty of knowing how individual actors or decision makers will respond to internal and external pressures for change. The last decade, and more particularly the last 5 years, has been a period in which agriculture and farmers have come under increasing pressure from a number of directions. Swinbank (1985) has suggested that price-support policies, investment subsidies, tax incentives and high inflation have resulted in a 'formidable squeeze on the need for a hired labour force', but the same measures help to explain the relative stability in the number of farmers. During the late 1980s further factors have continued to tighten the squeeze, including the tendency of the Common Agricultural Policy (CAP) to create structural production surpluses. As a consequence there has been pressure on farmers to reduce levels of production. Quotas on production were introduced for dairying in 1984, and threats about input or output quotas for arable cropping are regularly featured in the farming press. Suggestions now emerging from the European Commission (EC) indicate that CAP policies may be developed to encourage farming on a smaller scale, with perhaps a return to dividing up large holdings among members of a farm family. Such notions clearly run counter to the arguments in favour of scale economies which have held sway for so long.

Other pressures on farmers have arisen from the general public in two main directions. The public have become more concerned about the quality of the food they eat rather than just its price. Since the Second World War both Conservative and Labour governments have adopted a 'cheap food' policy: a rise in food prices is viewed as a vote loser. The demand for organically produced food, which has been associated with a concern over food quality, is volatile and relatively elastic. This is evidenced by the way in which the major food retailers have 'blown hot and cold' in their attempts to satisfy the demand. The second area of concern among the general public relates to the effects of modern farming practices on the environment. Increases in the level of car ownership and leisure time have encouraged people to regard the countryside as a location for recreation (see Chapter 11). This has led to a greater awareness that the environment of the countryside is not as portrayed in the popular image.

The emphasis in the mid to late 1980s shifted from production to diversification and survival, as income levels from agriculture fell. Agricultural policy still emphasizes efficient production but, in the presentation of policy, more stress is now being placed on what the consumer wants from agriculture in respect of methods of production, food quality and the impact of agriculture on the countryside. Indeed restructuring of the agricultural industry through the CAP has become a priority on the political agenda. Policy measures to take land out of agriculture were introduced in the late 1980s, such as setaside and incentives for forestry in lowland arable areas, but have generally received modest levels of support from the industry. A revised version of the forestry scheme, with more money available and lower area thresholds than the Farm Woodland Scheme, is now being considered by the industry. Initial indications suggest that it will be regarded more favourably. In 1991 there was an extension of development control, under the Town and Country Planning Legislation, to farms under 10 ha, which represents a significant departure from the privileged position that agriculture had enjoyed previously.

The linking of discussions about the CAP with *The Future of Rural Society* (European Commission, 1988) demonstrates that agricultural change does not occur in isolation; it is part of wider social and economic developments. The CAP has been at the heart of the European Community and its institutions. The transformation of this policy, and the development of an integrated policy for rural areas, reflects the need to accommodate the transition to an economic and social system in which farming is not a major employer of the national workforce. Agricultural policy is slowly being converted into a farm policy and funding from agricultural budgets is now being used to help alleviate rural problems. Whether such a change in policy can intervene to control the reduction in the farm labour force remains to be seen. It is still speculative to suggest that policies encouraging a more diversified and extensive agriculture will halt, let alone reverse, the long-term trend for decline in agricultural employment. It remains to be seen whether traditional forms of farming being fostered by such devices as Environmentally Sensitive Areas will lead to new employment opportunities in the industry.

References

Ball, R. (1986) Casual workers. *Agricultural Manpower* 2, 12.

Britton, D.K. (1968) Agricultural manpower: the current situation. In: National Economic Development Office, *Agricultural Manpower*. Proceedings of a Symposium at Manchester University, Manchester.

Champion, A.G. (1981) Population trends in rural Britain. *Population Trends* 26, 20–3.

Champion, A.G. (1982) *Counterurbanisation and Rural Rejuvenation in Britain: An Evaluation of Population Trends Since 1971*. Seminar Paper 38, Newcastle University Geography Department, Newcastle upon Tyne.

Champion, A.G., Green, A.E., Owen, D.W., Ellin, D.J. and Coombes, M.G. (1987) *Changing Places: Britain's Demographic, Economic and Social Complexion*. Edward Arnold, London.

Cloke, P.J. (1987) Rurality and change: some cautionary notes. *Journal of Rural Studies* 3, 71–6.

Cloke, P.J. and Edwards, G. (1986) Rurality in England and Wales 1981: a replication of the 1971 index. *Regional Studies* 20, 289–306.

Errington, A. (1988) Disguised unemployment in British agriculture. *Journal of Rural Studies* 4, 1–7.

Errington, A. (1990) Investigating rural employment in England. *Journal of Rural Studies* 6, 67–84.

European Commission (1988) *The Future of Rural Society*. Bulletin of the European Communities Supplement 4/88, Office for Official Publications of the European Community, Luxembourg.

Fielding, A. (1982) Counterurbanisation in Western Europe. *Progress in Planning* 17, 1–15.

Gasson, R. (1984) Farm women in Europe: their need for off-farm employment. *Sociologia Ruralis* 24, 222–7.

Grafton, D.J. (1982) Net migration, outmigration and remote rural areas. *Area* 14, 313–18.

Hodge, I. and Monk, S. (1991) *In Search of a Rural Economy*. Department of Land Economy, University of Cambridge.

Hoggart, K. (1990) Let's do away with rural. *Journal of Rural Studies* 6, 245–57.

HMSO (1991) *Regional Trends 1990*. Vol. 25. HMSO, London.

MAFF (annual) *Agricultural Statistics – England and Wales*. HMSO, London.

Lawton, R. (1968) Population changes in England and Wales in the later nineteenth century: an analysis of trends by registration districts. *Transactions of the Institute of British Geographers* 44, 55–74.

Little, J. (1991) Theoretical issues of women's non-agricultural employment in rural areas, with illustrations from the UK. *Journal of Rural Studies* 7, 99–105.

Munton, R., Marsden, T. and Whatmore, S. (1990) Agricultural restructuring: current trends and prospects. In: ACORA, *Faith in the Countryside*. Churchman and ACORA, Worthing and London.

Newby, H. (1977) *The Deferential Worker: a Study of Farm Workers in East Anglia*. Allen Lane, London.

Phillips, D. and Williams, A. (1984) *Rural Britain: a Social Geography*. Blackwell, Oxford.

Saville, J. (1957) *Rural Depopulation in England and Wales, 1851–1951*. London, Routledge.

Swinbank, A. (1985) A note on price support policy and hired farm labour. *Journal of Agricultural Economics* 36, 259–61.

Symes, D. (1991) Changing gender roles in productionist and post-productionist capitalist agriculture. *Journal of Rural Studies* 7, 85–90.

Walford, N.S. (1981) *The Development and Significance of Alternative Strategies in Agricultural Labour and Machinery Use*. Geography Research Paper 8, University of Sussex.

Walford, N.S. (1991) New faces in the countryside: social and economic restructuring in rural Britain. In: Brunet, P. (ed.) *Rural Britain and France*. Centre de Publications de L'Université de Caen, Caen, pp. 309–24.

Walford, N.S. and Hockey, A.E. (1991) *Social and Economic Restructuring in Rural Britain: a Methodology for Contextual Analysis*. Working Paper 18, ESRC Changing Countryside Initiative, University College, London.

13

PLURIACTIVITY AMONG FARM FAMILIES: SOME WEST EUROPEAN, US AND CANADIAN COMPARISONS

Anthony Fuller and Ray Bollman

Global economic restructuring is transforming the geopolitical map and labour relations of western industrial societies. The combined effect of the internationalization of capital, labour and product markets is a quantum leap in the process of change. Whether driven by technological, economic or environmental imperatives, restructuring has effects at the national, regional and local levels of society (Drucker, 1986; Kolko, 1988; Rosenau, 1988). New expressions of behaviour, both responsive and resistant, have emerged, which recognize and accommodate these imperatives and which have geographical expression and significance. One of these is the restructuring of rural labour markets (Newby, 1987).

Labour markets are traditionally tied to economic sectors and are often measured as spatial entities. Rural labour markets are being transformed as primary industries decline (or go 'offshore') and manufacturing becomes subject to the international division of labour. Global sourcing of the food sector, for example, means that the capital, labour, production and consumption involved in a single commodity can take place on four different continents. Labour skills and employment patterns are changing to short-term, part-time and multiple job-holding forms of work. Women's increased participation in the labour force is a factor, as are the processes of individualization and professionalization of social relations.

Labour markets have strong geographical meaning as they are primarily ordered over space. International finance, intellectual capital and materials can be transferred rapidly over space but labour markets are more fixed. Capital and production components come to labour markets which themselves are being continuously transformed by the state through training, research, migration and labour policies. The geographical separation of high-quality from low-quality labour markets is important, as is the differentiation within geographically defined labour sheds. The labour market approach is complex in that there are questions of gender, equity, class and social provision on the one hand and structural employment trends, such as the growth in the service sector, part-time work and de-skilling, on the other.

Restructuring in Agriculture

The restructuring of agriculture has produced two trends of importance for this debate. One is the current de-emphasis of primary food production and a shift to food manufacturing (Friedmann and McMichael, 1989). This has now been accompanied by pressure to reduce the support levels for farm commodities, the introduction of environmental controls or incentives and other measures to reduce agricultural production, while maintaining a competitive edge in agriculture through capital-intensive (technological) specialization (Warley, 1990). In this restructuring process, farm families are 'decoupled' from what had become traditional modes of support, that is, subsidies and guarantees tied to production. As in the past, they are encouraged to diversify, leave agriculture or turn to other uses of farm and family resources (Bollman and Smith, 1988; Ahearn *et al.*, 1990; Bollman, 1991; Bryden, 1991; Fuller, 1991).

The second outcome is the realization that rural development is potentially more sustainable than traditional agricultural policy, and certainly much cheaper (Commission of the European Communities, 1988; Freshwater and Ehrensaft, 1991; Aspen Institute, forthcoming). Restructuring of agricultural agencies is also taking place and we have seen a growth of interest in local economic development, rural employment strategies, community level programming and rural diversification in general (Fuller *et al.*, 1990).

These trends within agriculture in industrial nations may be summarized as shifts from:

Agricultural	to	Rural policy
Food from farming	to	Food from manufacturing
Family farms	to	Commodity production units
Mixed economies	to	Specialized economies
Farm Families	to	Land-holding families

Restructuring and Pluriactivity

One of the elements that captures the essence of the agricultural restructuring debate at the local level is the growth of pluriactivity. This is the propensity of farm households to engage in work other than farming, a mode of household economy that has reached major proportions in some parts of industrial society (Fuller, 1990). Although seemingly limited to the local scale, pluriactivity does transcend sectoral boundaries which challenges our conventional thinking about agrarian capitalism. More importantly, its ubiquitous presence in agrarian structures across Europe, the US and Canada raises questions about the suitability of agricultural versus rural policy.

There are two approaches to explaining the growth of pluriactivity in western farming:

1. Agricultural restructuring: the conventional paradigm argues that the continuation of the relative rise in the price of labour and the resulting substitution of technologically productive capital have increased the size of a productive unit that can provide full employment to a farmer (and the farmer's family). For the same number of families to remain on the land, farm families must find alternative sources of employment and income.

2. An alternative paradigm argues that social restructuring is altering concepts of work, unemployment and 'quality of life' values and goals. Women are entering the workforce in vast numbers, part-time work is becoming the norm, as is multiple job-holding, time-sharing and volunteerism. As these shifts become norms, it is not unreasonable to expect that farm households will adopt similar work patterns to achieve commonly held goals that emphasize individualism, self-employment and entrepreneurship. Such trends are visible in the restructuring of rural labour markets where more and more farm people find both work and social relevance.

With these two lines of argument in mind (both products of 'restructuring'), the evidence of farm family pluriactivity in the US, Canada and Europe is examined in order to verify its presence, identify some basic characteristics, and dispel some old myths. These two 'logics' are then revisited to see which element of restructuring best informs policy, whether agricultural or rural.

Pluriactivity – incidence

The incidence of pluriactivity can be measured in a variety of ways; but to make international comparisons possible, only data on the non-farm work of operators and spouses have been assembled here from secondary sources. The incidence of farm households in North America in 1987–88 with at least an operator or spouse with off-farm work was approximately 50% of all farm households (Table 13.1). As the countries are ranked, it is evident that the incidence of off-farm work by either the operator or spouse is higher in North America than for most West European countries, although the occurrence of spouses working off-farm is high in Ireland (53%) and Denmark (40%).

Table 13.1. Ranking of North American and European countries: off-farm work.

Country	% of operators with off-farm work	Country	% of spouses with off-farm work
US (1988b)	56	Ireland	53
Canada	48	Canada	50
US (1988a)	47	Denmark	40
US (1987)	43	US (1988b)	39
Germany	43	US (1988a)	33
Portugal	39	United Kingdom	29
Ireland	36	France	27
France	36	EUROPE 12	16
Greece	33	Italy	15
Belgium	33	Germany	8
Denmark	33	Belgium	8
EUROPE 12	30	Greece	7
Spain	28	Luxembourg	6
Italy	24	Netherlands	1
United Kingdom	24	US (1987)	n.a.
Netherlands	23	Portugal	n.a.
Luxembourg	18	Spain	n.a.

Source: Canada, 1986 Agriculture-Population Linkage; US, 1987 Census of Agriculture; US, 1988a Farm Costs and Returns Survey; US, 1988b Agricultural Economics and Land Ownership Survey; EUR10, 1985 Farm Structures Survey; EUR12, 1988 Agricultural Income Survey.

Although these figures are quite crude, an important myth can nevertheless be dispelled. It is evident that pluriactivity, whether measured in operator or spousal involvement with off-farm work, is more common in North America than in Europe where it has long been considered a more integral part of farming. The much vaunted family farm system of North America is clearly heavily involved with pluriactivity.

Pluriactivity – characteristics

One way to test if pluriactivity is a product of the restructuring of agriculture is to look for an association between farm size and the incidence of pluriactivity. Farm size can be usefully measured using the Standard Gross Margin (SGM) as a measure of business size. The SGM is a partial value-added measure. It is calculated as gross revenue or gross value of production minus selected expenses. European farms are classified according to Economic Size Units (ESUs) which equalled 1000 European Currency Units (ECUs) up to 1985 and 1100 ECUs in 1985. This chapter assumes 1 ESU = 1000 ECUs, and 1 ECU = 1 US dollar = 1 Canadian dollar. A strong correlation between farm (business) size and pluriactivity would suggest that pluriactivity arises (almost) solely from changes in farm structure. Restructuring would be expected to polarize the farm structure into a few large farms and many small farms whose families would participate in the off-farm labour market.

Table 13.2. Ranking of North American and European countries: standard gross margin (SGM).

Country	% of farms under 2000 SGM	Country	% of farms over 40 000 SGM
Italy	47	Netherlands	45
Greece	44	Canada	36
US	35	United Kingdom	29
EUROPE 10	35	Denmark	26
Ireland	33	US	25
United Kingdom	22	Belgium	19
Canada	18	France	15
Germany	17	Germany	11
Belgium	16	Luxembourg	11
France	15	EUROPE 10	9
Luxembourg	14	Ireland	3
Denmark	0	Italy	3
Netherlands	0	Greece	0

Source: Ahearn *et al.* (1990).

As expected, European farm sizes, when measured by standard gross margins, are smaller than those in North America (Table 13.2). Italy and Greece have the highest proportion of farms with less than 2000 SGM. At the other end of the scale, The Netherlands and Canada have the highest share of farms with over 40 000 SGM. Based on the restructuring hypothesis, one would expect that the relative farm size structure within a country would be strongly associated with the proportion of farm families allocating labour to off-farm labour markets. Lorenz curves provide a relative measure of structure. A Lorenz curve is constructed by ranking farms from smallest to largest along the horizontal axis and measuring on the vertical axis the cumulative proportion of aggregate SGM contributed by these farms. A Lorenz curve of the concentration of SGM indicates whether the farms within a country are relatively the same size (i.e. a low degree of concentration is depicted by a Lorenz curve near the diagonal) or whether there is a large proportion of small farms and a high degree of concentration of production on a few large farms. If the Lorenz curves suggest a relatively highly concentrated farm structure and a high proportion of relatively small farms, then a high proportion of operators and/ or spouses with off-farm work can be expected.

Table 13.3. Ranking of North American and European countries: concentration[a] of standard gross margin (SGM).

Most concentrated:	US[b]
	Italy
	United Kingdom, EUROPE 10, US[c]
	Ireland, Canada[b]
	Greece
	Germany, Canada[c]
	France, Belgium
	Luxembourg
	Denmark
Least concentrated:	Netherlands

[a]Countries are ranked by degree of concentration by visual comparison of Lorenz curves.
[b]Ranking when farms with SGM less than or equal to zero are *included*.
[c]Ranking when farms with SGM less than or equal to zero are *excluded*.
Source: Ahearn *et al.* (1990).

In relative terms, as portrayed by a Lorenz curve, the US farm structure was the most concentrated, if the 21% of farms with SGM less than or equal to zero are included. Otherwise, Italy has the most concentrated farm structure and The Netherlands the least concentrated (Table 13.3).

Overall, there appears to be no apparent relationship between the proportion of operators reporting some off-farm work and the proportion of small farms in the agricultural structure when comparisons are made across or among countries (Fig. 13.1). Neither is there an apparent relationship between the proportion of spouses reporting some off-farm work and the proportion of small farms in the agricultural structure of a country (Fig. 13.2). Italy, ranked as the country with the highest proportion of farms under 2000 SGM, with a Lorenz curve that shows a concentrated size structure, is ranked low in terms of the proportion of operators and spouses reporting some off-farm work (Figs 13.1 and 13.2). In general, pluriactivity is not associated with differences in farm size among countries.

However, there is an important association between the propensity to work off-farm and farm size (SGM) within countries. Operators of larger farms are associated with a lower participation rate in off-farm work in Canada, the US and Europe 10 (Fig. 13.3), and this holds true for each of the European countries. It is especially important to note that the participation in off-farm work by spouses is essentially flat across the farm size spectrum (Fig. 13.4). This differentiation in participation of operators and spouses in off-farm work according to farm size reflects a differentiation of gender roles in the farm household in relation to pluriactivity. Part-time farming research in the past has tended to ignore the role of farm women. Even at this very general level, a significant difference

Table 13.4. Distribution of farm operator households by major income source[a] and gross margin class.

Area and SGM class	Major income source (% of households)			
	Earned on-farm	Earned off-farm	Other income	Total
Canada, 1986	27	50	23	100
<$CAN 4000	5	70	25	100
$CAN 4000–15 999	11	62	27	100
$CAN 16 000 or more	42	37	21	100
US, 1986	23	55	22	100
<$US 4000	3	68	29	100
$US 4000–15 999	13	61	26	100
$US 16 000 or more	50	37	13	100
EC Study Areas, 1987	52	28	20	100
<4 ESU[b]	24	42	33	100
4–16 ESU[b]	57	25	17	100
16 or more ESU[b]	84	10	6	100

[a]Major income source is defined as 50% or more. Households receiving 50% or more of their income from unearned sources or not receiving 50% from any one source are included in the other income category.
[b]European Size Unit.
Sources: Canada, Statistics Canada, 1986 Agriculture-Population Linkage; US, 1986 Farm Costs and Returns Survey; EC Study Areas, 1987 Arkleton Trust Baseline Survey.

Anthony Fuller and Ray Bollman

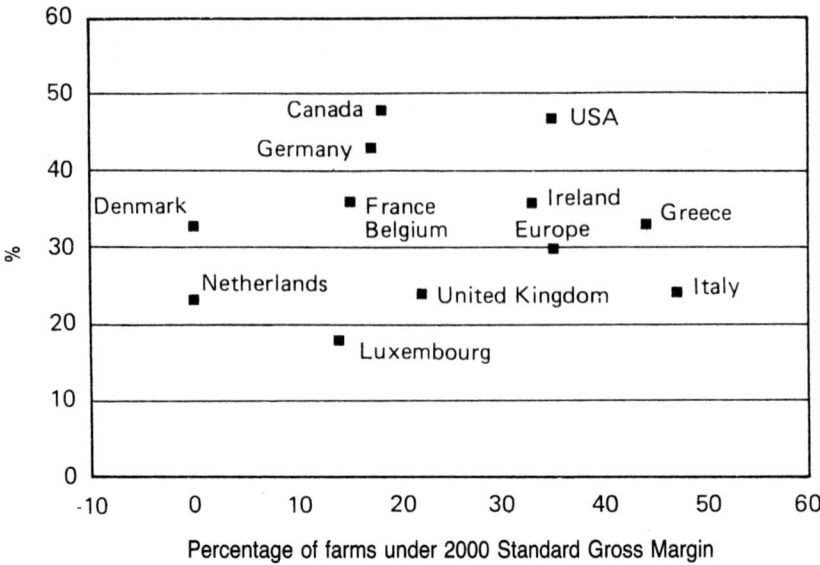

Fig. 13.1. Off-farm work by operators.

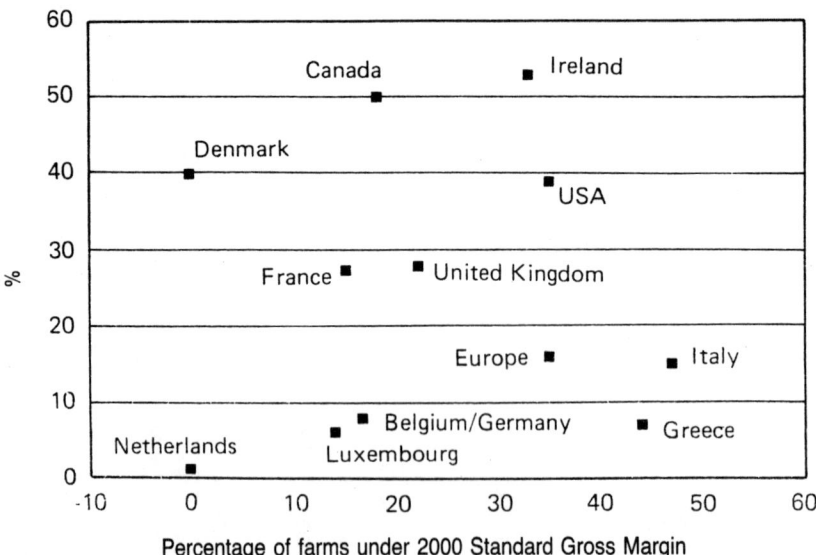

Fig. 13.2. Off-farm work by spouses.

appears to characterize pluriactivity according to gender across farm size classes within countries.

A final comment on the characteristics of pluriactivity in western industrial farming society reflects the significance of off-farm incomes in the total income structure of farm families. Income earned off-farm is important, especially in the US and Canada where in all farm size categories it is over 37% of total income (Table 13.4). In the US and Canada over half of farm households gain more than 50% of their total income from off-farm work or from other income, such as social transfers, remittances and return on investments. This is irrespective of farm size, although there are noticeably more households earning 50% or more of their total income from farm earnings among households on larger farms (SGM). In Western Europe, the study area data from the Arkleton Trust Farm Household Survey in 1987 indicate that the dependency on non-farm sources of income is substantially lower.

Conclusions

From this cursory glimpse of the macro-level data it is possible to appreciate the significance of off-farm dependence by farm people, both in terms of work and incomes. When on-farm diversification activities are added, the picture is even more one of interdependency between farm communities and local and regional labour markets. The importance of rural labour markets for farm family survival, growth and reproduction cannot be ignored, although there are surprisingly few studies of what farm people do in their off-farm work and what, in fact, constitutes a farming community's labour market.

Given the importance of rural and regional markets for the survival and development of farm populations, it is crucial to understand better the changing structures of such markets and how they themselves are being affected by global restructuring. Whether farm people participate in the well-remunerated part of the labour markets (young farm women frequently worked as teachers and nurses in the past) or are being drawn into low-paid, part-time work is yet to be clearly established. The Arkleton Trust surveys of 24 study areas in 12 countries of Western Europe will illuminate some of these questions, but much more needs to be done (Arkleton Trust, 1989; MacKinnon *et al.*, 1990).

This brings us back to the question of pluriactivity and restructuring. There is ample evidence to suggest that its growth and importance is not simply the result of the restructuring of agriculture. There are only weak associations between pluriactivity and farm size (SGM), farm type, land tenure and capitalization. Pluriactivity appears to be especially associated with social restructuring, being the result of women entering the paid

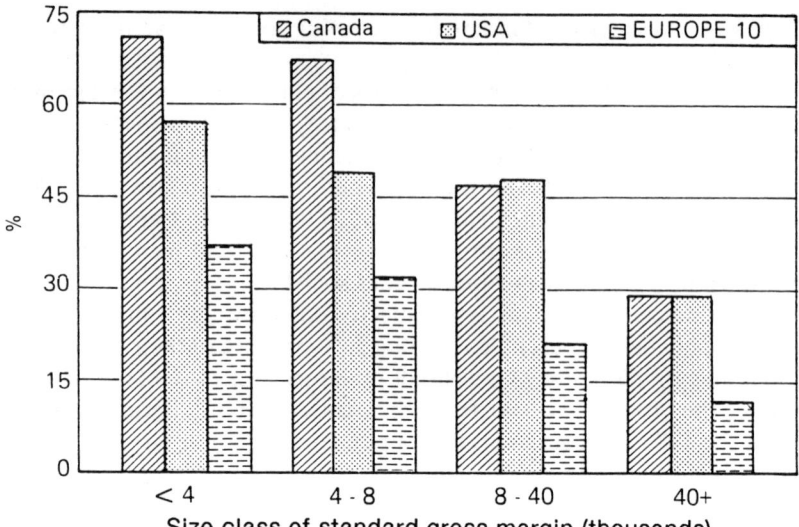

Europe 10 has lowest participation
in off−farm work across all size classes

Fig. 13.3. Operators with off-farm work.

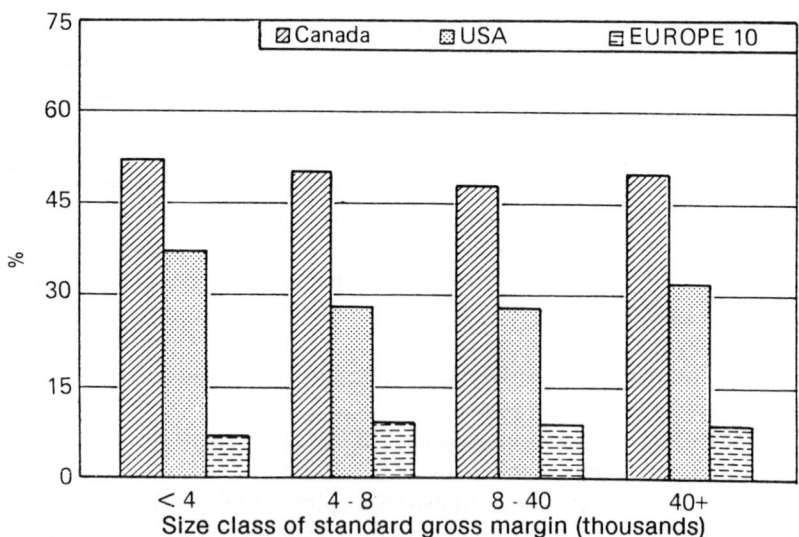

Europe 10 has lowest participation
in off−farm work across all size classes

Fig. 13.4. Spouses with off-farm work.

labour force (off-farm), the growth in preference for part-time work and the decline of single-career paths. It seems that because we organize our affairs in sectors and administrative divisions, and because restructuring of the global economy is most visible (at the outset) at the sector level, we subscribe to the structural myopia of the past. Interpreting pluriactivity in strictly agricultural terms is clearly a limited and false exercise. Pluriactivity is an old phenomenon (Bollman, 1982), but is larger in scale and scope now, reflecting societal change and the economic reality of many rural families. While agricultural policy-makers struggle to get their sums right and return agriculture to a full-time farming basis, many farm families have solved the problem for themselves, although not without many new stresses and unexpected strains. It is in this light that greater attention to rural development policy would seem to have merit. Studies which reveal how rural and regional labour markets work, and how farm people participate in them, would be as effective for the public good as the endless 'cost of production' studies of farm commodities. Researchers too will need to cross boundaries, disciplinary boundaries, to examine the nature and future prospects of rural and regional labour markets in an age of global restructuring.

References

Ahearn, M., Bollman, R. and Fuller, A.M. (1990) *The Role of Pluriactivity in Maintaining Incomes of Farm Households – Comparisons of Canada, United States and the European Community*. Paper presented at AAEA/CAEFMS Conference, University of British Columbia.

Arkleton Trust (1989) *First Report for the Commission of the European Communities on Structural Change and the Use Made of Structures Policies by Farm Households in the European Community*. Arkleton Trust (Research) Ltd, Langholm.

Aspen Institute (forthcoming) *Conceptual Frameworks for Understanding Rural Development: An International Dialogue*. The Aspen Institute, Washington DC.

Bollman, R. (1982) Part-time farming in Canada: issues and non-issues. *GeoJournal* 6, 313–22.

Bollman, R. (1991) Efficiency aspects of part-time farming. In: Hallberg, M. *et al.* (eds), *Multiple Job-Holding among Farm Families*. Iowa State University Press, Ames, pp. 112–39.

Bollman, R. and Smith, P. (1988) *Integration of Canadian Farm and Off-farm Markets and the Off-Farm Work of Women, Men and Children*. Research Paper Series 16, Statistics Canada, Ottawa.

Bryden, J. (1991) The impact of 1992 on rural Europe. In: Tracy, M. (ed.) *Rural Policy Issues*. Arkleton Trust (Research) Ltd., Longholm, pp. 13–26.

Commission of the European Communities (1988) *The Future of Rural Society.* COM, 501, Final, Brussels.

Drucker, P.F. (1986) The changed world economy. *Foreign Affairs*, March, pp. 768–91.

Freshwater, D. and Ehrensaft, P. (1991) Rural policy development in the USA and Canada: issues and prospects. In: Tracy, M. (ed.) *Rural Policy Issues.* Arkleton Trust (Research) Ltd, Langholm, pp. 85–106.

Friedmann, H. and McMichael, P. (1989) Agriculture and the state system: the rise and decline of national agricultures, 1870 to the present. *Sociologia Ruralis* 2, 93–118.

Fuller, A.M. (1990) From part-time farming to pluriactivity: a decade of change in rural Europe. *Journal of Rural Studies* 6, 361–73.

Fuller, A.M. (1991) Multiple job holding among farm families in Canada. In: Hallberg, M. *et al.* (eds) *Multiple Job-Holding among Farm Families.* Iowa State University Press, Ames, pp. 31–44.

Fuller, A., Ehrensaft, P. and Gertler, G. (1990) Sustainable rural communities in Canada: issues and prospects. In: Gertler, M. and Baker, H. (eds) *Sustainable Rural Communities in Canada.* Proceedings of the ARRG Rural Policy Seminar, Department of Extension, University of Saskatchewan, Saskatoon, pp. 1–41.

Kolko, J. (1988) *Restructuring the World Economy.* Pantheon, New York.

MacKinnon, N., Bryden, J., Bell, C., Fuller, A.M. and Spearman, M. (1990) Pluriactivity, structural change and farm household vulnerability in Western Europe. *Sociologia Ruralis* 31, 58–71.

Newby, H. (1987) Economic restructuring and rural labour markets in Europe: current policy options. In: Summers, G. (ed.) *Agriculture and Beyond, Rural Economic Development.* University of Wisconsin Press, Madison, pp. 41–54.

Rosenau, J.N. (1988) Patterned chaos in global life: structure and process in the two worlds of world politics. *International Political Science Review* 9, 327–64.

Warley, T.K. (1990) Megatrends affecting agrifood and rural society. *Canadian Journal of Agricultural Economics* 38, 717–25.

14

THE SUSTAINABILITY OF PRAIRIE RURAL COMMUNITIES

John Everitt and Robert Annis

The rural areas of the Prairie Provinces have, like other non-urban areas of Canada, been suffering through a series of crises in recent years of both an economic and a sociocultural nature. Although nearly 80% of the Canadian population is now urban, major regions of the country – such as the sparsely populated prairies – are still characterized by more traditional patterns of rural community living. In Manitoba, for instance, over 40% of the population is widely scattered in the province's rural areas, including 74 incorporated settlements, and nearly 500 additional unincorporated settlements (Everitt and Gill, forthcoming).

These towns and villages have been the mainstay of the prairie farm society for most of their existence (Friesen, 1974), but many are currently in serious economic difficulty with between 40% and 50% of the incorporated Prairie settlements declining in numbers (Annis, 1989; Davies, 1990; Everitt and Gill, forthcoming). This reflects a decrease in rural population in general, and farm population in particular, which is one facet of the major changes that are occurring due to the new economics of agriculture in this region. An often overlooked consequence of this transformation is that the present quality of life in the rural areas, social as well as economic, is threatened, as is the potential life-style for future generations. That is to say, rural areas are becoming less 'sustainable' and less 'healthy' (Everitt et al., 1990; Everitt and Bessant, 1992).

Neither the rural dweller, nor the provincial and federal governments, seem to have much control over these events at present, but all must learn to counter their effects – or risk a further erosion of rural Prairie life. Although efforts have been made at the local and provincial level, it is the federal level from which most energy and funding has been forthcoming up until the present. Unfortunately, though commonly involving overt government intervention into the rural economic subsystem, these responses, while numerous, have mostly been limited and ineffective (Everitt and Bessant, 1992), and there is a general lack of institutional or governmental structures that could be used to address rural problems in an integrative fashion (Corbett, 1988; Annis, 1990; Fuller *et al.*, 1990). Although recently there has been some evidence that this is entering into the Canadian political agenda, the country's typical response has been one of *laissez-faire* with some 'crisis management' thrown in where necessary.

Why Should Rural Communities be Supported?

Major parts of the Canadian economy and society have always been based upon agricultural land, and upon the rural communities that live on this land (Fuller *et al.*, 1990; Everitt and Gill, forthcoming). This chapter is primarily concerned with the sustainability of small communities on the Prairies, and particularly with those in Manitoba. But the issues have a wider significance: they can be generalized, in many instances, to other parts of Canada where established life-styles have also been recently eroded. Seven major reasons can be identified why more attention should be paid to the maintenance and enhancement of these life-styles.

Perhaps the most important reason is that, in a multicultural country and world, the attempt should be made to retain, promote and sustain as many alternative life-styles as possible. This is at least an ethical argument, but also it seems to be only common sense to keep the options open. In a country that promotes multiculturalism and bilingualism, it seems only right to preserve a valid alternative for future generations.

Second, an argument can be made for the validity of the 'small rural community' life-style. There should be a viable alternative to large urban centres, and the smaller community has the perceived advantage of what many feel is a better life-style, with a closer-knit community, despite a number of counteracting disadvantages related to housing, transportation and services.

Third, currently there exists a large economic investment in an existing rural infrastructure. If this investment deteriorates and more people move to the cities, then new infrastructures will have to be built in the larger urban areas, and this is bound to be a more expensive option.

Fourth, there may well be some ecological benefits to retaining rural communities. One such benefit could lie in the better use of fossil fuels, as people living in small rural communities should have a proportionately greater amount of more efficient driving, and their use of petrol should be more productive. Rural communities also have the capability of being more self-sufficient in terms of food production, and thus of becoming an integral part of a 'sustainable society' (Brown *et al.*, 1990).

Fifth, if the Prairies were one large depopulated agricultural machine, with just a few large cities, it would be easier for a few people, or corporations, to threaten the environment as there would be no one out there to 'look over it', and guard and sustain it. This is not to suggest that Prairie farmers have never threatened and degraded their environments over the past century, nor to imply that all decisions made in far-off boardrooms are ecologically dangerous, but rather it reflects a fear that the situation could well deteriorate if the 'owners' of the land did not live on it.

Sixth, rural communities should be sustained in order to protect the food supply. Canadian consumers are insufficiently aware that their meat, bread, pasta and other foods come from Prairie farms, even though it is made very clear by the supermarkets that fruit and other produce comes from California, Texas or Florida. If farmers are to be retained on the land to produce this food, they need to be provided with community services, such as schools, churches and social services, as well as groceries and tractor parts.

Seventh, economy and society exist as a system, and the actions that take place in one part of the system potentially affect all the other parts. Even if urban dwellers do not 'use' small communities, they need them in order to make the system work at its most efficient level.

Thus there are several reasons for sustaining rural communities. However, under the present economic systems of Canada and the world, the sustainability of these small communities will not be guaranteed without some short-term financial cost. Some might argue that if the 'reasons' for survival are sufficiently valued by Canadians, then the 'market' will lead to the 'provision' of small communities. However, this market is not so free as urban-based economists might suggest; rather, the 'economic deck' is currently 'stacked' against the small town. Thus, not only should the government intervene, but it should intervene more equitably.

To bring about change in the current urban-rural interaction processes (to 'unstack the deck'), a redistribution of financial resources will undoubtedly be necessary. But the 'majority' will need to look out for and protect the interests of the 'minority'. On a purely financial basis, no investment of funds in such 'sustainability' is one option, while a 'bottomless pit' lies at the other end of the continuum. Where does the appropriate level of investment lie? How can the economic costs be reconciled with the advantages of maintaining the integrity of small communities? If it is an ethical question and not a purely financial one, the answer is easy, but life is never this simple. It is up to Canadian society to decide where the fulcrum between ethics and bottom-line economics must lie.

What Are the Threats to Rural Communities?

The survival of rural areas, and communities, is a battle fought not on one front but on many. A major threat comes from the fact that the

Prairies are in many ways like a developing nation, acting as a hinterland producing primary products and controlled by the heartlands of Central Canada, Europe and the United States (Fuller *et al.*, 1990). Recent battles over such matters as railway abandonment, railway passenger service reduction, postal service and free trade have made it clear that the Prairies are dominated by external forces, which if not countered by a concerted effort from the 'grass roots', as well as the higher levels of government (i.e. both 'bottom-up' and 'top-down'), will lead to the further decline and destruction of the small community. Rural development is, for instance, very clearly tied to agricultural policies, commodity prices, multi-national corporate profits, and external 'efficiencies'. These features do not incorporate non-economic viewpoints very well, if at all, such as the value of supporting the rural community for its own sake, or for the sake of its inhabitants.

In a Canadian context, there has been little research undertaken to explain the particular vulnerability of many small settlements to change which 'undermines or destroys the community's *raison d'être* and eventually its actual existence' (Douglas, 1990, p. 67). The paucity of theory in this area is lamentable, but a beginning has been made by Douglas (1989, 1990). He suggests that there are three interconnected propositions (a 'nexus') that must be faced and dealt with in order to give small rural settlements a chance at sustainability. This nexus has to be viewed in the light of the 'new reality' of rural life which exhibits such characteristics as:

1. an increase in the price of land, in average farm size coupled with a decrease in the number of farms, and a profound change in land ownership patterns;
2. a reduction in the farm population as a proportion of the total rural population, coupled with a reduction in the farm labour force;
3. an increase in the need for off-farm work in order to maintain the family farm, and thus an increasing dependence of farms upon the local communities;
4. an increase in the proportion of farm families with relatively low incomes, and particularly low incomes from agricultural sources;
5. an increase in the proportion of seniors in the population both due to our ageing population within the country as a whole, and to the flight of younger people from smaller communities;
6. a decrease (or at least a smaller increase) in rural services, such as schools and physicians, when compared with their urban counterparts.

Douglas' (1990, p. 69) first proposition suggests that there is an extensive diffusion of scarce internal energies within a community, which is closely intertwined with the 'new reality' of rural life. On occasion this 'diffusion' can lead to an uncritical acceptance of some information sets and apparent opportunities, and can mean that longer-term priorities are prejudiced in favour of apparent shorter-term gains. This problem reflects both the lack of trained personnel at the disposal of rural communities, as well as their inability to communicate quickly with other communities that might find themselves in similar situations.

Second, Douglas (1990) points out that some of the fundamental characteristics of 'rurality', many of which have been exacerbated by the 'new reality', militate against the viability of rural communities. He summarizes these characteristics as an 'impedance to mass', by which he means that the lack of a dense population causes rural communities to be vulnerable to a number of potentially dangerous outside influences. The challenge for rural communities is to gain mass through communications (especially horizontal), new coalitions, intermunicipal collaborative structures, innovative networks and so on.

Third, Douglas (1990, p. 70) believes that a condition (or culture) of sustained dependency exists 'which is both internally as well as externally nurtured'. This not only detracts from the efficacy of indigenous or internal initiatives, but constantly places the community in a reactive posture responding to Federal, Provincial or Territorial government programmes and priorities. Further, vulnerability is exacerbated not only by exposure to the vagaries of so-called senior government funding priorities, but by the inattention to and the consequent creeping erosion of indigenous entrepreneurship; as a consequence, the capacity to develop community-responsive, or customized initiatives is abridged, thereby jeopardizing any attempts at sustainability, yet alone viability.

Whither Rural Communities?

The social, demographic and economic changes occurring in the Prairie Provinces are complex: they are a consequence of a large number of interrelated factors, not of any single influence. This is what makes small community issues so difficult to understand and resolve, and to project the outcome of these trends into the next century is an even more complicated exercise. However, some points are clear. For instance, although most rural communities owed their genesis to, and were dependent upon, agriculture (Brierley and Todd, 1990), to a large extent this dependency has been reversed, and the viability of many farms is dependent upon the economic health of nearby communities. Given this kind of rural reversal, what are some of the factors that may help sustain the small community?

One key feature is Local involvement in any initiative; this involvement would include human resources, financial resources and ideas, information and technology. As Howell has noted:

the vast majority of communities have within themselves and within their grasp a considerable capacity to develop. It is attitude that ultimately makes a difference . . . I have never seen a rural area, a cluster of rural countries, or a city in America that failed because of excessive boldness.

(Howell, 1983, pp. 23–4)

But even then change will most likely be slow: instant transfusions of success are few and far between.

Another feature is that there has to be a better marriage between 'top-down' and 'bottom-up' approaches. Rural community sustainability needs some external help, but more critically needs collective rather than individual responses at the local level. It follows that there is not and should not be a 'blueprint' that is applied in some uniform manner to all communities. The 'top-down' approach too commonly adopts such a fallacy and leads to the failure of otherwise promising programmes.

It is clear that the rural system is not developing in an unproblematical way. Therefore, what can be done about this and what kind of actions may make a difference in a long-term effort to sustain rural communities? Table 14.1 summarizes what community residents can do in order to work towards survival.

One of the major problems related to the sustainability of small com-

Table 14.1. Rural community survival checklist.

Does your community have:

1. Evidence of community pride and spirit?
2. A willingness to invest time, talents and money in the future of the community?
3. Active participation in decisions affecting the community?
4. A willingness to transfer power to younger leaders?
5. An acceptance of women in leadership roles?
6. A sufficient supply of leaders, or leaders-in-training?
7. A strong belief in the value of education?
8. Media that encourage diversity of opinion?
9. Good access to information and new technology?
10. A well-maintained infrastructure?
11. An active economic development programme?
12. A willingness to cooperate with neighbouring communities?
13. A willingness to seek advice from the outside?
14. A programme that encourages young people to return to the community following post-secondary training?
15. Sufficient cultural and recreational activities?
16. A belief that self-help is the best help for both individuals and communities?
17. A continuing interest in new ideas for development?

Source: Annis (1992).

munities is the lack of an adequate data base to permit precise understanding of exactly what is happening to our small towns. More research based on local case studies is needed, and this needs to involve local people in the research process. Federal and provincial policies are required that are not based on the assumption that rural communities are going to die. For this, concepts of political and economic feasibility have to be redefined. As Ruckelshaus (1989, p.174) indicates (in the context of the 'ecological world'), 'these concepts are, after all, simply human constructs; they were different in the past, and they will surely change in the future'. Such a redefinition should allow for local initiatives within a regional framework, and will undoubtedly cost money. However, if the higher levels of government were convinced of the need to help rural communities, this could cut down on the depopulation that empties the small towns, and forces the large urban areas to expend capital on increasing their basic installations and facilities.

Although the entire nation has a stake in the welfare of rural communities, members of these communities have to invest their surplus monies in their own future: economic development is also a local responsibility. Communities need to avoid a 'low tax ideology'; they must be willing to raise taxes both to gain money and to gain a sense of empowerment. This would also enable the preservation of the rural infrastructure and discourage further out-migration. Direct public subsidies should be limited, visible and well targeted. Participants must have a direct stake in a programme's success.

Rural communities must come to define 'community' broadly and gain 'mass' through enhanced communications, new coalitions, inter-municipal collaborative structures, innovative networks and so on. Rural communities should view each other as cooperative neighbours rather than as competitors. Horizontal and vertical networks have to be developed for channelling resources and information in order to obtain an integration of effort. That is to say, communities need to develop effective processes for responding in an efficient, focused manner to opportunities and threats. Such processes should clearly include both planning and implementation. What are currently lacking are 'forms of communication that connect the overarching interest of society . . . with the well-being of the individual' rural resident (Crosson and Rosenberg, 1989, p. 135).

Flexible leadership, an increased turnover in public office, and the sharing of informal roles, in order to 'spread the load', have to be encouraged so as to take advantage of all possible energies, and help to marry the 'top down' and 'bottom up' approaches to development. Services have to be maintained and improved, including education and access to medical facilities within small communities, in order to increase their attractiveness. A strategy is needed to retain and utilize the benefits of highly skilled people in the rural areas. In order to break the chain of dependency, it may be useful to encourage sources of information of all sorts – which may, on occasion, be controversial. The acceptance of controversy and a debate over alternatives may thus be seen as acceptable and normal, rather than something to be avoided. A continuing forum for debate should be encouraged in rural communities. Sustainable rural communities must be forward-looking communities that empower the local people.

Finally, each person tends to cling tenaciously to an inalienable right to live where (s)he wants, while failing to relate the repercussions of that individual decision to regional and national socioeconomic trends. Each separate decision may be good for the individual but lead to a collective calamity. Thus Canadian society must decide whether or not to preserve rural communities. Some people suggest that 'market forces' should take the decision and that a better form of rural life will rise 'Phoenix-like' from the ashes of the present. However, it is hard to believe this variation on the old Vietnam War adage, 'it is necessary to destroy rural life in order to save it'.

It can be seen that most of the solutions presented in this chapter are 'social' rather than 'economic' and/or technical. Many of the past attempts at 'rural revival' were the exact opposite, but the sustainability of small communities is very much a social – and a political – problem. It is a problem of persuading the decision makers at all levels to give adequate consideration to the value of the small community over the 'long haul', and that the expansion of the economy as a whole need not be at the expense of small communities. It is also, of course, a problem of educating the (majority) urban voting public of this necessity.

Conclusion

Given the arguments presented in this chapter, it should not be surprising that a question frequently raised with respect to rural communities is 'why do they persist?' In many ways the contemporary Prairie settlement has exhibited little soundness or vitality, and the immediate future promises little better. These rural dwellers could probably do better, economically, elsewhere. At present they survive as a result of unorthodox work patterns, government financial programmes and, conceivably, simple inertia. Perhaps a more revealing issue is what Corbett (1988) has referred to as the 'perceptual paradox': why do so many small communities not only persist but grow, given the commonly expounded disadvantages of living in them?

As indicated earlier, over 40% of small Prairie settlements are getting smaller. Clearly over 50% are getting larger. Although the growth of individual settlements can be understood by proximity to larger urban centres, by the provision of water and other services, by changes in job opportunities, and by the transportation network, these explanations seem to consist of necessary but insufficient conditions. An interpretation beyond that of economic viability is clearly needed and that seems to lie in the subjective judgement of 'quality of life'. The occupants of these communities like living where they are, are prepared to stay when economic conditions are decidedly 'unhealthy', and seem likely to continue this pattern into the foreseeable future.

In addition, however, some communities are attempting to improve their life-styles and change their chances of survival, from the 'bottom up'. In many parts of the Prairies there are community leaders talking about more local jobs, and more local improvements. These individuals are among that growing number who have reached a major conclusion: salvation begins at home. For them it has become quite clear that neither the federal nor the provincial governments can solve all the local problems. These higher levels of government may sponsor some programmes of assistance, but generally these are of the type that are triggered by a local initiative. Consequently the mayors, and reeves, and chamber presidents who once looked outward for new payrolls, now realize that they must look at themselves, at their neighbours, and at their friends and associates. It is likely that the great majority of all new jobs to be created in Prairie towns will arise from local initiative. It also seems clear that attempts to foster the growth and economic development of small towns and rural regions will probably flounder, unless they are based upon the realization that subjective ('socio') factors may be as important to development as are the more familiar objective ('economic') criteria. The sustainability of Prairie rural communities is a complex issue, and its achievement will not be accomplished without the dedication and commitment of local Prairie town initiatives.

References

Annis, R.C. (1989) *Strategic Planning for Rural Development*. Westarc Group Inc., Brandon University, Brandon.

Annis, R.C. (1990) Maintaining Alberta's rural population: a challenge to planning in the '90s. In: Oldman River Regional Planning Commission, *Proceedings of the 31st Annual Workshop on Rural Depopulation: How will it affect our lives?* The Commission, Lethbridge, pp. 34–46.

Annis, R.C. (1992) *A Look at Communities Most Likely to Succeed*. Westarc Group Inc., Brandon University, Brandon.

Brierley, J.S. and Todd, D. (1990) *Prairie Small-Town Survival: The Challenge of Agro-Manitoba*. Edwin Mellen Press, Lampeter.

Brown, L.R., Flavin, C. and Postal, S. (1990) Picturing a sustainable society. In: Worldwatch Institute, *State of the World 1990*. Report on Progress Toward a Sustainable Society. W.W. Norton and Company, New York, pp. 173–90.

Corbett, R. (1988) Between the devil and the deep blue Atlantic: the dilemma of rural and small town development in Atlantic Canada. A paper presented to the Canadian Urban and Housing Studies Conference, Winnipeg, Manitoba.

Crosson, P.R. and Rosenberg, N.J. (1989) Strategies for agriculture. *Scientific American* September, 128–35.

Davies, W.K.D. (1990) What population turnaround?: some Canadian Prairie settlement perspectives, 1971–1986. *Geoforum* 21, 303–20.

Douglas, D.J.A. (1989) Community economic development in rural Canada: a critical overview. *Plan Canada* 29, 28–46.

Douglas, D.J.A. (1990) Rural community development and sustainability. In: Gertler, M.E. and Baker, H.R. (eds) *Sustainable Rural Communities in Canada*. The Canadian Agriculture and Rural Restructuring Group, Saskatoon, Saskatchewan, pp. 66–72.

Everitt, J.C., Annis, R.C. and McGuinness, F. (1990) The responsibility of urban dwellers to foster sustainable rural communities. In: Beavis, M.A. (ed.) *Ethical Dimension of Sustainable Development and Urbanization: Seminar Papers*. Institute of Urban Studies, Occasional Paper, Winnipeg, pp. 119–48.

Everitt, J.C. and Bessant, K. (1992) Regional planning in rural Manitoba. *Applied Geography* 12, 65–79.

Everitt, J.C. and Gill, A. (forthcoming) The social geography of the small Canadian town. In: Ley, D. and Bourne, L. (eds) *The Social Geography of Canadian Cities*. McGill-Queen's University Press, Kingston and Montreal.

Friesen, G. (1974) *The Canadian Prairies*. University of Toronto Press, Toronto.

Fuller, A.M., Ehrensaft, P. and Gertler, M. (1990) Sustainable rural communities in Canada: issues and prospects. In: Gertler, M.E. and Baker, H.R. (eds), *Sustainable Rural Communities in Canada*. The Canadian Agriculture and Rural Restructuring Group, Saskatoon, Saskatchewan, pp. 1–41.

Howell, J.M. (1983) Economic development of the future: myths and realities. In: North Central Regional Center for Rural Development, *Proceedings of the Community Economic Development Strategies Conference*. The Center, Iowa State University, Ames, Iowa, pp. 23–4.

Ruckelshaus, W.D. (1989) Toward a sustainable world. *Scientific American* September, 166–74.

DEVELOPMENT STRATEGIES
FOR RURAL COMMUNITIES

15

ECONOMIC DEVELOPMENT VERSUS LAND USE REGULATION

Paul Frederic

Economic development versus environmental protection has been the focus of much of the discussion concerning the future of America (Friedman, 1979; Étzioni, 1983; Farr, 1983; Mandelker, 1986; Lapping *et al.*, 1989). The alleged added costs of meeting environmental regulations is viewed by some public policy makers and developers as an unnecessary burden to economic growth. This issue, when combined with the facts that: (i) rural regions tend to lag behind urban places in most measures of fiscal well-being; and (ii) much of the environmental movement is driven by urban people who value the countryside as a place to play rather than to make a living, represents a major challenge for rural communities. Given the set of development and environmental programmes directed toward non-metropolitan America, can the economic gap between it and high population density areas be closed?

Over the past two decades most rural regions and small towns have continued to trail metropolitan communities in economic opportunities (Redman and Rowley, 1990). During this same period a wide variety of rural-oriented economic and environmental protection programmes have been enacted by individual states. The latter is usually in the form of land use controls. This chapter examines the relationship between non-metropolitan and metropolitan economic trends relative to development incentives and environmental land use regulation.

Development Programmes and Regulation Trends

Federal and state roles in rural policy have changed over time. Until the early 1970s national rural development efforts focused on farm policy; however, more recent shifts in the economic structure of the countryside have resulted in efforts to encourage non-farm growth through infrastructure improvement, housing development and business loans (Lapping *et al.*, 1989, pp. 16–17). Relative growth or slower decline in the non-natural resource sectors of rural economies suggests that non-metropolitan systems are becoming more complex. Hart (1988) argues that manufacturing is displacing agriculture as the major element for economic growth in the rural Middle West, and Mitchelson and Fisher (1987) found that an expanding commuting range was a major factor in rural transformation. Non-employment income represents a growing portion of all income in some areas of rural America (Manson and Groop, 1988; Thomas *et al.*, 1989). In addition to these three trends, Bonner (1986, p. 341) notes that many people prefer to live in small towns and the open country. In conclusion, rural development is associated with: (i) increased manufacturing; (ii) extended commuting; (iii) non-employment income, and (iv) life-style preference.

This change in the rural system occurred during a time when the federal government moved resources away from development programmes and states began to invest more in solving the problems of small towns and low population density areas. However, in places where state resources did not replace reduced federal funds, local communities found it difficult to maintain traditional levels of services and economic support programmes. Small agriculturally dependent towns were especially hard hit (Koven and Koven, 1989).

By 1988, 35 different types of state rural development programmes were in place (Hackett, 1988, p. 434). These included: (i) agriculture; related development; (ii) transition tools to assist in movement out of agriculture; (iii) rural business assistance, and (iv) rural community assistance (Table 15.1). This diversity of programmes indicates that states have been innovative in addressing their individual needs. The character of rural places varies from region to region. For example, most people in the northeast live within commuting distance of a significant urban employment centre. On the other hand, it is often a long drive to the nearest off-farm job in the Great Plains. Programmes that might fit high growth, compact, industrial New Hampshire, could be a poor choice for slow growth, agricultural South Dakota.

In addition to the above collection of development programmes, rural areas have been the target of many environmental land use regulations (Table 15.2). Included are farmland preservation (Furuseth and Pierce, 1982; Nelson, 1988), forest practices (Brown, 1986; Henly and Effefson, 1986; Vale, 1988), wetland protection statutes (Marcelli, 1980; Dubensky, 1990; Williams, 1991) and broad land-use controls (Lapping *et al.*, 1989, pp. 72–103; Rohse, 1987). Objectives and effectiveness are uncertain with many of these regulations. Furuseth (1985) found that some combination of level of agriculture, government expenditures and taxation, population increase, size of minority groups and political orientation suggests which California counties have enacted farmland protection laws. However,

Table 15.1. Types of state rural development programmes.

I. *Agriculture-related development*
    1. Agricultural export development
    2. Attracting value-adding business
    3. Beginning-farmer programme
    4. Biotechnology and technological transfer
    5. Crop diversification
    6. Marketing agricultural products
    7. Other

II. *Transition tools*
    1. Agricultural and rural development commissions, agencies, etc.
    2. Assessing competitive advantages
    3. Farmer and agribusiness financial programmes
    4. Farmer retraining and counselling
    5. Other

III. *Rural business assistance*
    1. Economic development comprehensive
    2. Entrepreneurship, business incubators, etc.
    3. Job creation training
    4. Location of new business/industry
    5. Marketing and export
    6. Plant/military base closing
    7. Procurement assistance
    8. Retention and expansion of existing business
    9. Small business assistance
    10. Rural enterprises zones
    11. Tax incentive for private investment
    12. Technology transfers
    13. Other

IV. *Rural community assistance*
    1. Culture and arts
    2. Financial
    3. Housing
    4. Infrastructure
    5. Land use
    6. Parks and recreation
    7. Quality of life
    8. Tourism
    9. Day care services
    10. Other

Source: Hackett (1988) Extract reprinted from *Book of the States* 27, p. 343.

these broad findings are of limited utility in targeting programmes to meet individual state or regional needs. A nationwide review by Nelson (1988, pp. 106–7) concludes that purchase of development rights is the best way

Table 15.2. Types of rural-oriented environmental land use programmes.

I. *Farmland preservation*
    1. Differential tax assessment
    2. Circuit breaker taxation
    3. Central land use policies
    4. Agricultural land banking
    5. Transfer of development rights
    6. Exclusive agriculture zoning
    7. Agricultural districting
    8. Executive powers action
    9. Waiver of urban infrastructure assessment

II. *Forest practices*
    1. Timber production
    2. Water protection
    3. Clean air production
    4. Soil protection
    5. Wildlife habitat protection
    6. Protection/recreational opportunities

III. *Wetland and shoreland protection*
    1. Wetlands management
    2. Statewide shoreland action
    3. Permits required for activities affecting the resource
    4. Regulated under a federally approved coastal management plan
    5. State law authorized local units to adopt regulations pursuant to state standards
    6. Permits required under shorelands zoning regulations
    7. Vermont site plans review act established standards and requires approval for most resource-disturbing activities

IV. *Statewide land use management*
    1. State has authority to require permits for certain types of development
    2. State established mechanism to coordinate state land use-related problems
    3. State requires local governments to establish a mechanism for land use

Source: Furuseth and Pierce (1982), p. 197; Nelson (1988), pp. 84–5; Henly and Effefson (1986), p. 9; Maine (1991); Marcelli (1980), p. 515; Matthews (1976), p. 478; Kusler, J. (1983), pp. 130–4; Matthews (1976), p. 478.

to preserve farmland. He points out that other options only continue to subsidize speculators. Henly and Effefson (1986, p. 16) note that forest practice acts tend to have one of three primary objectives: (i) to regulate proliferating local laws; (ii) to ensure long-term wood-fibre production, and (iii) to protect soil, water and wildlife. A fourth catch-all group includes issues of timber theft, protecting land-owner rights and large-scale industrial operations that offend many newcomers to forested regions. Wetland programmes focus on water quality and wildlife habitat. During the 1980s, federal policy shifted from a passive wetland management format to a position of protection (Lapping *et al.*, 1989, p. 242). General land use regulations tend to deal with the conflict between economic change and environmental quality. Growth management is a common theme of land use control. The Maine Land Use Regulation Commission (Maine, 1983) represents a standard approach to comprehensive land use control. The agency's primary mission is to provide reasonable opportunities for development while protecting the natural environment.

This assemblage of regulatory laws has produced a complex net of statutes that are sometimes ineffective and the public often finds difficult to understand. Clark (1991) notes that the present federal plant-closing regulatory framework is not working. He points out that there are inconsistent guidelines and confusion is common among legislative practices, courts and public agencies. The author's experience as Director of the Maine Land Use Regulation Commission, from 1987 to 1989, confirms Clarke's argument. Despite efforts to avoid the problem, contradicting guidelines are often produced by the legislative bodies or through agency rule-making activity. Frequent statute change also adds to regulatory uncertainty. For example, the Maine State Legislature revised the definition of a subdivision pertaining to unincorporated areas six times during a 7-year period. This produced much consternation among both land owners and public agencies that regulate development. The relationship between the legislature and public agency charged with enforcing statutes must be well defined if the system is to work (Frederic, 1991).

Conceptual Framework

In the following analysis, rural economic development programmes (Econtot) and land use regulation patterns (Envitot) have been compared with non-metropolitan economic progress relative to metropolitan progress, in 48 of the 50 states (Table 15.3). Selected data on Alaska are missing, and New Jersey has no non-metropolitan population; thus, both are excluded from the study. United States Census Bureau guidelines are used to define non-metropolitan and metropolitan areas:

> An area qualifies for recognition as a metropolitan area (MA) if (1) it includes a city of at least 50,000 population, or (2) it includes a Census Bureau defined urbanized area of at least 50,000 within a total metropolitan population of at least 100,000 (75,000 in New England). In addition to the country containing the main city or urbanized area, an MA may include othercountries having strong commuting ties to the central county.
>
> (United States, 1989, p. 24)

Each state is assigned an economic development (Econtot) value based on the number of rural development programmes it has in place

Table 15.3. Economic gap, programmes and non-metropolitan population.

State	Gap non-metro-politan/ metropolitan jobs growth rate 1979–81[a]	Econtot[b]	Envitot[c]	Percentage of population that is non-metropolitan[d]
Alabama	2.7	23	1	35.6
Alaska				60.3
Arizona	−22.9	22	2	22.8
Arkansas	− 6.3	10	7	56.9
California	− 6.6	22	9	4.8
Colorado	−10.8	9	1	18.3
Connecticut	− 2.8	5	7	8.0
Delaware	− 0.7	14	5	33.8
Florida	− 3.6	9	7	9.8
Georgia	1.9	20	3	35.0
Hawaii	20.7	15	7	22.7
Idaho	−11.7	15	2	17.5
Illinois	− 8.6	29	2	31.8
Indiana	− 3.7	13	8	57.7
Iowa	− 7.4	27	1	80.9
Kansas	−16.2	12	1	19.0
Kentucky	− 4.9	23	1	54.6
Louisiana	− 7.6	24	3	33.3
Maine	− 4.4	20	12	59.5
Maryland	1.3	7	12	7.0
Massachusetts	21.9	11	7	4.3
Michigan	4.7	28	2	19.7
Minnesota	−15.1	23	8	34.0
Mississippi	−12.3	16	2	70.0
Missouri	− 3.4	18	3	33.9
Montana	0.1	23	5	76.5
Nebraska	−17.2	22	2	53.2
Nevada	1.1	17	6	16.8
New Hampshire	− 6.0	8	6	37.6
New Jersey				0.0
New Mexico	−22.2	18	4	59.0
New York	− 2.2	22	9	9.5
North Carolina	−15.2	11	4	44.4
North Dakota	−16.1	23	1	63.5
Ohio	− 3.7	10	1	21.0
Oklahoma	− 9.3	12	1	42.0
Oregon	− 6.7	18	12	32.3
Pennsylvania	− 3.8	17	4	15.3
Rhode Island	33.6	7	5	8.6
South Carolina	−11.1	10	1	39.2
South Dakota	−17.9	8	1	82.5

Table 15.3. Continued

State	Gap non-metro-politan/ metropolitan jobs growth rate 1979–81[a]	Econtot[b]	Envitot[c]	Percentage of population that is non-metropolitan[d]
Tennessee	−10.7	21	2	34.5
Texas	−14.9	9	3	19.1
Utah	− 5.6	18	2	22.8
Vermont	−13.4	17	8	75.1
Virginia	−20.9	22	8	27.9
Washington	−15.7	19	13	18.7
West Virginia	4.6	17	1	64.1
Wisconsin	− 1.6	23	8	33.3
Wyoming	22.6	20	3	89.0
All States (Av.)	− 5.1	16.8	4.6	36.6

Source: [a]Computed by author using data from Bureau of Economic Analysis, United States Department of Commerce, Washington, District of Columbia.
[b]Total number of rural development programmes. Compiled by author from: Hackett (1988), p. 434.
[c]Total number of environmental land-use programmes. Compiled by author from: Furuseth and Pierce (1982), p. 194; Nelson (1988), p. 86; Maine (1991); Marcelli (1980), p. 515; Matthews (1976), p. 478.
[d]Computed by author using data from Bureau of the Census, United States Department of Commerce, Washington, District of Columbia.

(Table 15.3). An environmental land use regulation (Envitot) value that reflects the total of farmland preservation, forest practices, wetland protection and general land use control laws in statutes is also generated. Rural economic well-being is measured in terms of employment growth in non-metropolitan versus metropolitan areas. The percentage of the population that is non-metropolitan is also compared with change in size of employment growth: this acts as a locational factor that is not development programme or environmental law driven. The 1979–87 period represents most of the 1980s and thus should be a good base from which to view the new decade of the 1990s.

Analysis

State-level rural economic development programmes are most numerous in the Northern Plains, Middle West and South (Fig. 15.1). All states with 23 or more programmes in place are part of a belt extending from the interior of Dixie through the industrial and agricultural heartland into the small grain and cattle land of the High Plains. On the other hand, state-

level rural oriented land use protection laws are more frequent in the
Northeast and the West Coast (Fig. 15.2). New York-Vermont-Maine,
and California-Oregon-Washington represent two distinct clusters. Most
of the states with extensive economic development support options have
passed few environmental land-use controls. Results of the Wilcoxon
signed-rank test suggest a strong negative relationship between the two
types of legislation (Table 15.4).

Table 15.4. Wilcoxon signed-rank test results.

Variables	Z-scores
Econtot–Envitot	−6.06
Gap in Employment Growth–Econtot	−5.76
Gap in Employment Growth–Envitot	4.56
Gap in Employment Growth–Percentage of Population Non-metropolitan	5.88

Source: Computed by author. All Z scores significant at 0.95 level of confidence.

One might argue that the classic pro-development versus pro-environ-
ment struggle has produced these extremes. However, non-metropolitan
versus metropolitan economic trends, as measured by the increase in
number of jobs in each area, indicates growth in most areas during the
8-year study period (Fig. 15.3). Only West Virginia experienced decline
in both rural and city jobs. Although new jobs are a useful measure of
growth, they do not show the entire picture. When adjusted for inflation,
earnings per job, with the exception of metropolitan California, declined
in all areas of the US other than along the Atlantic Seaboard. New
England and New York experienced an especially strong performance
(Redman and Rowley, 1990). Despite the overall loss in earnings per job
throughout most of the nation, increase in new jobs is a useful indicator
of how well non-metropolitan areas are doing compared with metropolitan
regions. Are rural places and small cities gaining or dropping further
behind compared with large urban communities?

In nine widely scattered states, non-metropolitan places created jobs
faster or lost them slower than metropolitan areas (Fig. 15.4). In either
case, large cities did poorer than other places. Non-metropolitan Massachusetts and Rhode Island (close to an oxymoron) were part of the Northeast regional economic boom of the 1980s and sustained rapid growth.
On the other hand, metropolitan Wyoming (another near oxymoron)
lost jobs while the rest of the state gained. Non-metropolitan economies
performed the weakest in the High Plains and Southwest. Depressed
agricultural conditions in the former hurt rural places more than large
communities. In the case of the Southwest, urban 'sunbelt' growth left
rural and small-town Arizona and New Mexico behind. Z values indicate a

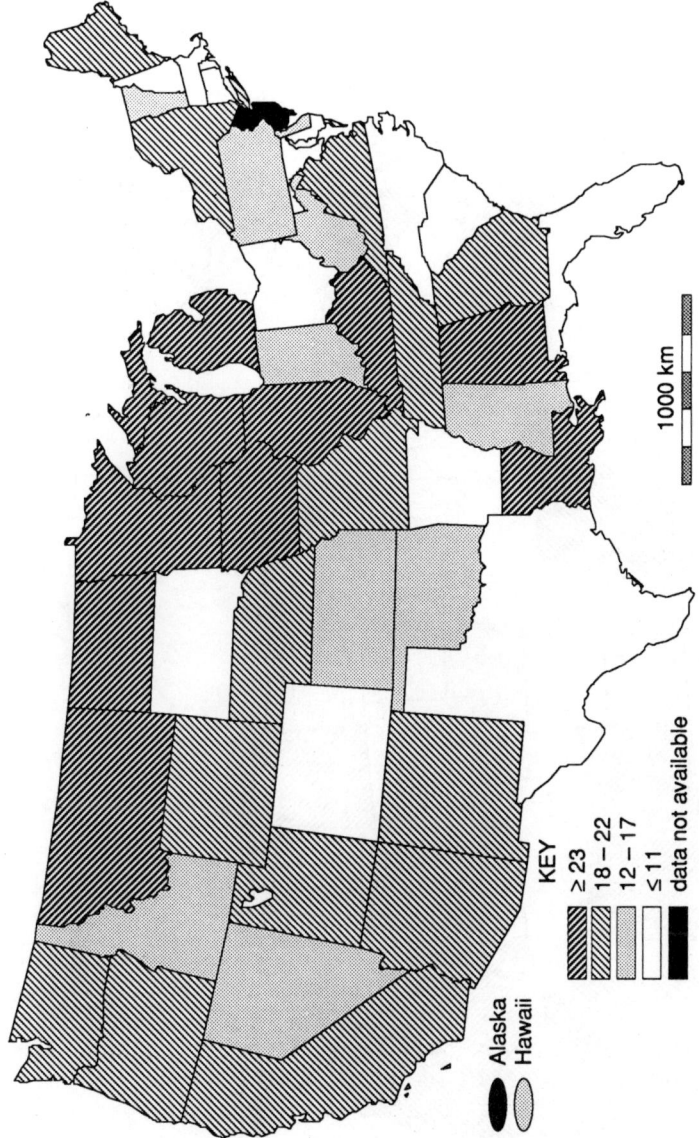

Fig. 15.1. Number of rural economy programmes by state.

KEY

≥ 23

18 – 22

12 – 17

≤ 11

data not available

Alaska

Hawaii

1000 km

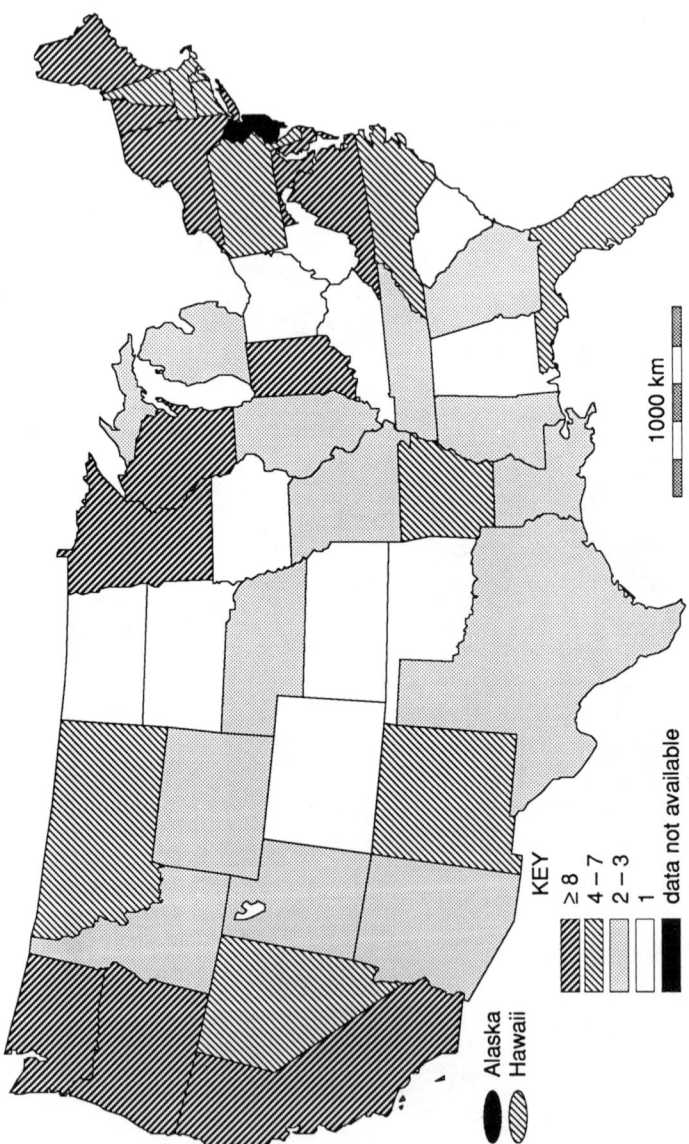

Fig. 15.2. Number of rural oriented land use regulations by state.

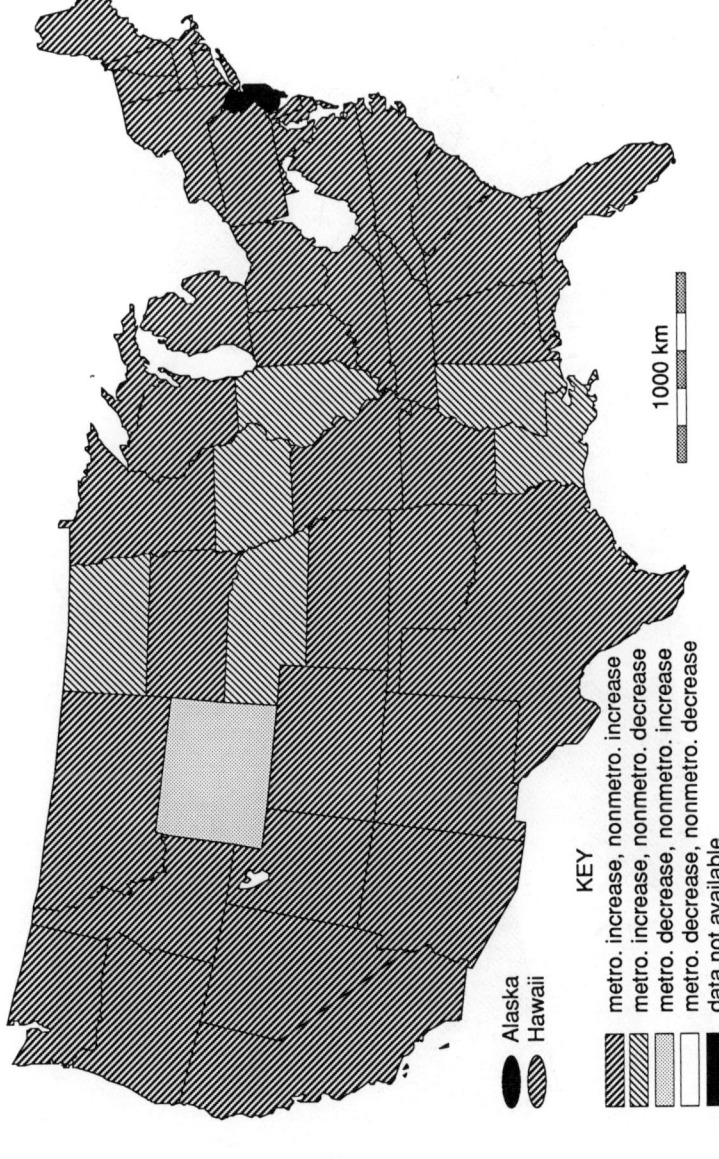

KEY

Alaska
Hawaii

metro. increase, nonmetro. increase
metro. increase, nonmetro. decrease
metro. decrease, nonmetro. increase
metro. decrease, nonmetro. decrease
data not available

1000 km

Fig. 15.3. Employment change, 1979–87.

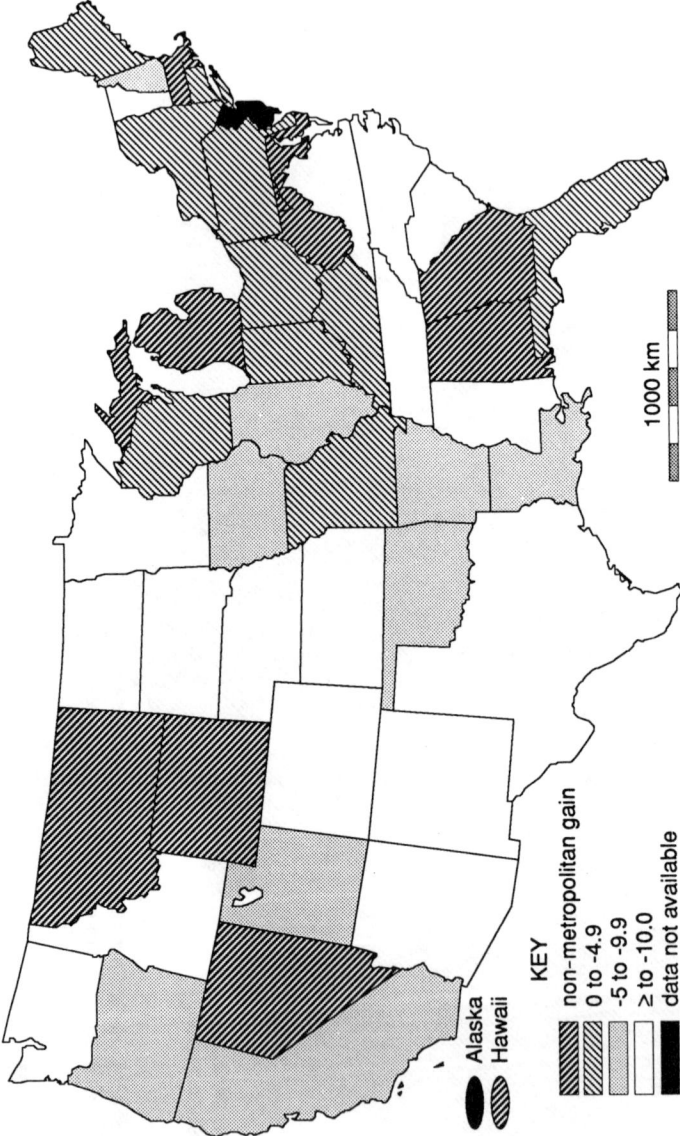

Fig. 15.4. Percentage change in gap between non-metropolitan and metropolitan, 1979–87.

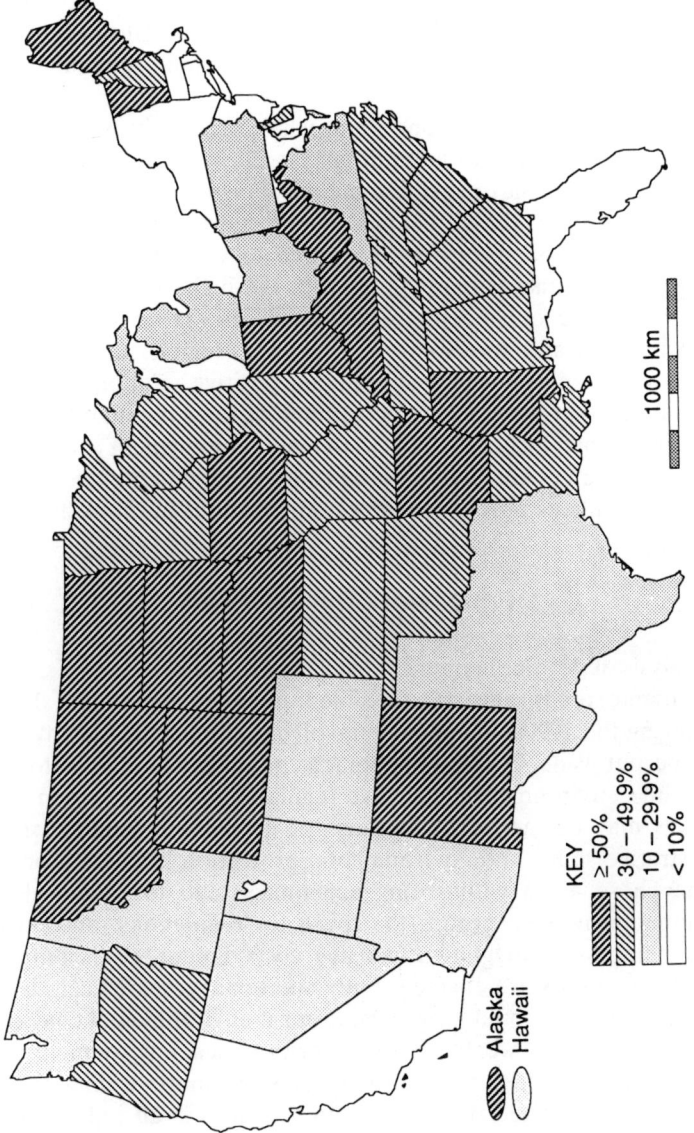

Fig. 15.5. Percentage of population that is non-metropolitan.

strong negative association between economic trends in non-metropolitan regions and the number of rural development programmes (Table 15.4). On the other hand, states with the best records in closing or slowing the widening of the non-metropolitan–metropolitan gap have the most comprehensive environmental land-use legislation. Business incentives have done little to boost rural development and land-use regulations have not suppressed it.

These results raise questions about what forces drive or depress rural development. Proximity to large urban places, and their associated economic opportunities, may provide some advantage to small towns and rural areas. Consequently the percentage of the state population that is non-metropolitan has been examined relative to economic performance in low population density areas. States with the highest proportion of their populations living outside metropolitan places are concentrated in the Northern Plains and the South (Fig. 15.5). Regions that have seen a widening of the employment gap have a positive relationship (Z score) with locations that have a high percentage of their population living outside of large urban locations. Access to major employment centres appears to have a strong positive influence on rural development, whereas neither economic incentives nor environmental regulations have much impact.

Framework for the 1990s

This analysis of development incentives, environmental controls, ruralness and economic trends suggests a possible framework to guide rural development during the 1990s. The debate involving jobs versus the environment should be put aside. Rural development programmes appear to do little good, while environmental land-use regulations do not prevent growth: incentives may be ineffective because of inadequate funding or lack of administrative skills, and environmental protection laws are a product of growth. Those areas with strong economies tend to have well-funded environmental interest groups that push for regulations, and businesses that are doing well enough to afford new costs. Under these circumstances new protective laws are enacted. Only additional research can determine if economic growth would have been faster if environmental controls were weaker. In regions of weak economic performance, environmental groups are not well supported and business may be more willing to combat regulations. Indeed the diversity of programmes is an indication of the power of various special interest groups. Swanson (1989) comments that past rural development policy has been fragmented among many special groups such as agriculture, recreation and newcomers to the countryside. This problem can be overcome if a comprehensive policy is formulated. Better connections between rural and urban regions are critical if the economic gap separating the two is to be reduced. Investment in linking infrastructure appears to be a useful strategy. However, caution should be the watchword as North American economic trends have taken a downturn during the early 1990s. Recent patterns indicate that New England, one of the strongest growth regions of the 1980s, may be experiencing a deeper recession than the rest of the United States.

Conclusions

Rural public policy has involved a fragmented approach driven by special interest groups. The result has been a variety of economic incentive programmes and environmental land-use laws. However, these have had little impact in narrowing or widening the economic gap between non-metropolitan and metropolitan places. This pattern indicates the jobs-versus-environment debate should be put aside relative to rural policy. On the other hand, nearness to metropolitan centres is a positive influence on rural economic growth.

Additional work should focus on linkage between rural and urban places. Both economic and environmental rural public policy should be better integrated with urban concerns. The most productive approach may be a linkage plan that involves both the country and city.

Acknowledgements

The author thanks Michael Leahey for assistance with data management, Stacey Hodges for cartographic work, and Elizabeth Gosselin for typing the manuscript.

References

Bonner, W. (1986) Rural development. In: So, F., Hand, I. and McDowell, B. (eds), *The Practice of State and Regional Planning*. American Planning Association, Chicago, pp. 340–68.

Brown, R. (1986) Environmental and natural resource problems: The role of the states. *Book of the States* 26, 401–19.

Clark, G. (1991) Limits of statutory response to corporate restructuring illustrated with reference to plant closing legislation. *Economic Geography* 67, 22–41.

Dubensky, M. (1990) Wetland regulations – beware the net. *Tree Farmer* (Summer), 12.

Etzioni, A. (1983) *An Immodest Agenda: Rebuilding America Before the Twenty-first Century*. McGraw-Hill, New York.

Farr, C. (1983) Modifying land use regulations for economic development. In:

Carr, J. and Duensing, E. (eds) *Land Use Issues of the 1980s.* Center for Urban Policy Research, Rutgers University, New Brunswick, pp. 207–28.

Frederic, P. (1991) Public policy and land development: the Maine Land Use Regulation Commission. *Land Use Policy* 8, 50–62.

Friedman, S. (1979) Economic development: the planning response. In: So, F., Stollman, I. and Beal, F. (eds) *The Practice of Local Government Planning.* International City Management Association, Washington DC, pp. 589–99.

Furuseth, O. (1985) Influences on county farmland protection efforts in California: a discriminant analysis. *Professional Geographer* 37, 443–51.

Furuseth, O. and Pierce, J. (1982) A comparative analysis of farmland preservation programmes in North America. *Canadian Geographer* 26, 191–206.

Hackett, J. (1988) Agricultural and rural development. *Book of the States* 27, 424–35.

Hart, J. (1988) Small towns and manufacturing. *Geographical Review* 78, 272–87.

Henly, R. and Effefson, P. (1986) *State Forest Practices Regulations in the United States: Administration, Cost and Accomplishment.* Station Bulletin AD–3011, Agricultural Experiment Station, University of Minnesota, St. Paul.

Koven, S. and Koven, A. (1989) Responding to fiscal constraints: management of small towns in the farmbelt. *Journal of Rural Studies* 5, 295–8.

Kusler, J. (1983) Regulating sensitive lands: an overview of programs. In: Carr, J. and Duensing, E. (eds) *Land Use Issues of the 1980s.* Center for Urban Policy Research, Rutgers University, New Brunswick, pp. 128–53.

Lapping, M., Daniels, T. and Keller, J. (1989) *Rural Planning and Development in the United States.* Guilford, New York.

Maine (1983) *Comprehensive Land Use Plan for Areas Within the Jurisdiction of the Maine Land Use Regulation Commission.* Main Land Use Regulation Commission, Maine Department of Conservation, Augusta, Maine.

Maine (1991) Unpublished files. Bureau of Forestry, Maine Department of Conservation, Augusta, Maine.

Mandelker, D. (1986) Environmental protection. In: So, F., Hand, I. and McDowell, B. (eds) *The Practice of State and Regional Planning.* American Planning Association, Chicago, pp. 480–99.

Manson, G. and Groop, R. (1988) Concentrations of nonemployment income in the United States. *Professional Geographer* 40, 444–50.

Marcelli, R. (1980) Pollution and the environment. *Book of the States* 23, 497–517.

Matthews, R. (1976) State land use programs. *Book of the States* 21, 474–8.

Mitchelson, R. and Fisher, J. (1987) Long distance commuting and income change in the towns of Upstate New York. *Economic Geography* 63, 48–65.

Nelson, A. (1988) An appraisal of farmland preservation policies: which work as intended? In: Corbett, R. (ed.) *Protecting Our Common Future: Conflict Resolution Within the Farming Community.* Rural and Small Town Research and Studies Programme, Mount Allison University, Sackville, pp. 79–109.

Redman, J. and Rowley, T. (1990) What your state's earnings growth says about the need for rural development policy. *Rural Development Perspectives* (February), 30–2.

Rohse, M. (1987) *Land Use Planning in Oregon.* Oregon State University Press, Corvallis.

Swanson, L. (1989) The rural development dilemma. *Resources* (Summer), 14–16.

Thomas, R., Sommers, L. and Kamiar, M. (1989) An index of rural development for Southwest Michigan: the role of nonemployment income. In: Gustafsson, G. (ed.) *Development in Marginal Areas*. Department of Geography, University of Karlstad, Karlstad, pp. 47–60.

United States (1989) *Residents of Farms and Rural Areas*. Series P-20, Number 446, Current Population Reports, Bureau of the Census, United States Government Printing Office, Washington DC.

Vale, T. (1988) Clearcut logging, vegetation dynamics and human wisdom. *Geographical Review* 78, 376–86.

Williams, M. (1991) The human use of wetlands. *Progress in Human Geography* 15, 1–22.

16

FEDERAL WATER RESOURCE DEVELOPMENT IN THE UNITED STATES

Lizbeth Pyle

Sustainable development has begun to compete with simple economic growth as a primary objective in rural areas. The goal of sustainable development requires that three competing objectives be incorporated in efforts to improve the quality of life in rural communities: (i) promoting economic growth through diversification; (ii) accommodating social needs of the population, and (iii) maintaining the integrity of the environment (Armstrong-Cummings, 1986; Smit and Brklacich, 1989). In the past, people expressed far less concern for the degree to which the environment was sacrificed in efforts to expand employment and income. Especially in the years following the Depression, the goal of economic growth took precedence. One favoured plan for spurring employment, income and, ultimately, regional development that emerged as part of Roosevelt's New Deal in the 1930s, centred on federally funded water projects (Clawson, 1981). The federal government reasoned that water projects would put Americans to work during construction, and eventually in industries lured to the area by the improvements.

Legislators traded votes as favours to one another in garnering projects for their constituents. They authorized the construction of many dams that currently interrupt the flow of rivers across the country. Heavy federal subsidies of capital, operational and maintenance costs for water resource projects result in considerable transfer of wealth, so projects have been viewed as a 'plum' for areas in which they have been constructed (Wahl, 1989). In recent years, few projects have been undertaken because the huge national debt can no longer support the proliferation of such heavily subsidized projects. As Congress cut back on capital-intensive development projects, community leaders have struggled to identify alternatives to boost their local economies.

This chapter reviews the anticipated link of federal water projects to local economic development. A brief review of the major federal agencies charged with water resource development is followed by a critique of the methods and assumptions used in evaluating water projects, especially cost–benefit accounting. Finally, suggestions are made relative to the need that reorganization of water resource management accompany the restructuring of rural employment, as places pursue sustainable development.

Major Federal Water Development Agencies

Two federal agencies have dominated water resource development: the US Army Corps of Engineers and the US Bureau of Reclamation. The Tennessee Valley Authority (TVA), a third agency with regional scope, also developed a large number of federal water projects. Each of these agencies began their history of water development with a specific mandate.

From 1933 to 1943, the National Resources Planning Board coordinated their development activities (Clawson, 1981). Though short-lived, the board laid the foundation for water planning and management in the United States and no single organization assumed all the functions it performed. The process became fragmented, with duplication of some efforts and oversight of others.

Over time, the roles of the Corps, the Bureau, and the TVA broadened to include multiple agendas. Projects providing flood control, power generation and recreation, for example, could be marketed more effectively than those with a single purpose. Multipurpose dam and reservoir complexes increased in popularity, especially during the 1950s and 1960s.

The US Army Corps of Engineers' long association with water projects began in 1824, when it assumed responsibilities for river and harbour improvements (Holmes, 1972). Most early projects involved dredging and other navigation improvements. Since that time, it has become geographically and functionally one of the most diversified federal water resource managers, assuming responsibilities for flood control, dam safety and some wetlands. Most of the large reservoirs in the eastern United States, as well as a smaller number in the West, were constructed and are still managed by the Corps.

The Bureau of Reclamation, on the other hand, is exclusively a western agency. The *Reclamation Act* of 1902 established the Bureau to provide and manage water projects that would facilitate settlement of the arid West (Warne, 1973). The Bureau met its mandate so successfully that it essentially has worked itself out of a job. In recent years the agency has moved from water resource development to water resource management (Moore, 1991). Like the Corps of Engineers, many of the Bureau's projects have been marketed as multipurpose reservoirs. They provide reservoir-based recreation opportunities and power generation, as well as irrigation or flood control.

The Tennessee Valley Authority is unique in that it exercises federal functions within a single region. The TVA was created as part of Roose-velt's New Deal in 1933 to provide flood control and improve navigation (Holmes, 1972). Government officials also hoped the TVA would stimu-late regional economic development, land and water conservation, and rural electrification. It became the nation's largest utility, in part because of the government's frustration with the private electrical utility companies (Clawson, 1981). The agency continues to generate electrical power and its reservoirs still provide flood control and recreation, but this New Deal experiment was not repeated in other regions of the country.

Evaluation of Federal Water Projects

While the costs of federally funded water projects are borne by taxpayers across the country, benefits accrue to particular regions. Congress obvi-ously is reluctant to authorize expenditures unless the benefits to society will outweigh the costs incurred. National economic efficiency has been measured using cost-benefit analysis. This technique is used to calculate a ratio of all project costs to all project benefits; benefits must exceed costs for projects to receive favourable reviews. Congress has acknowledged the technique in principle since the New Deal, but it has guided water project evaluation explicitly only since the 1950s (Krutilla and Eckstein, 1958). In recent decades, cost-benefit analysis has become the premier technique used to evaluate a wide range of programmes, not just water development.

Despite the explicit goal of efficiency, the extremely political nature of water project authorization cannot be overemphasized (Bromley and Barrows, 1974). Many projects originated as unfavorable (yet weakly justified) reports to Congress, but were modified repeatedly until they could be authorized. Human values other than efficiency have also affected public work decisions. Krutilla and Eckstein (1958, p. 265) acknowledged that 'society has used water resources as a means of providing employment and settlement opportunities, of fostering the growth of underdeveloped or depressed regions, and promoting the widespread use of electric power'. Despite this laudable national goal, the effectiveness of such strategies has been questioned.

Employment issues

Cost-benefit analysis typically includes secondary benefits, not just direct ones such as flood protection (Holmes, 1972). It is here that the employ-ment issue surfaces. The employment of individuals in the construction of water projects and in private enterprise, stimulated as a result of a water project, has been viewed as the most important secondary benefit. Such benefits can be used to help cancel-out labour and other construction costs in trying to calculate the positive cost-benefit ratio required to justify a project.

Accurate estimates of secondary benefits are difficult to derive, especially in areas of unemployment or underemployment where water projects are seen as a means of directly improving the local economy. In communities with less than full employment, the cost of employing otherwise idle (or underemployed) workers would be less than the industry-wide wage rate. Tolley (1966) emphasized the need to refine techniques for estimating costs and benefits of projects in depressed areas, especially when assuming that labour mobility and effective demand result in full employment (Haveman and Krutilla, 1968). Those with a more humanistic perspective recognize that attachment to place and other human ties keep labour from being perfectly mobile.

Water projects frequently have been justified as good employment generators. Haveman and Krutilla (1968, pp. 87–8) suggested that water projects had a high employment ratio per dollar of expenditures when compared with labour compensation as a percentage of the Gross National Product (GNP) (72% for multipurpose water projects, 56% for the GNP). Even though a higher percentage of money has been expended on wages for those employed in water projects than the percentage of money expended on wages for the GNP as a whole, the employment benefits of federally funded water projects typically only last during construction phases. Perhaps more importantly, water projects have created only a modest demand for unskilled labour, so chronically depressed areas have benefitted little from such federal expenditures (Haveman and Krutilla, 1968). Projects have brought skilled labourers and managers to their sites, instead of trying to fill all positions from within the local labour pool.

Furthermore, federal water projects have not built a strong record of sustaining employment and regional development. A number of case studies concluded that the availability of water or water projects alone was insufficient to cause economic growth; water projects may have facilitated, but did not 'cause' growth (Butcher, 1974). Another study emphasized the role of entrepreneurship, and other social connections, in regional development over low-cost power or navigation improvements provided by water projects (Chandler, 1984, pp. 61–2).

Value of recreational and scenic amenities

Federal water projects have drawn criticism for reasons other than questionable employment benefits. A frequent complaint has been the low discount rate employed in estimating the present value of future costs and benefits associated with projects (Berkman and Viscusi, 1973; Palmer, 1986; Reisner, 1986). Low interest rates insufficiently emphasize the needs and desires of present generations in favour of future benefits from the primary purpose of water projects (usually flood control or irrigation). Water development agencies have neglected the value that present generations place on streams for recreation or preservation.

Agencies are being challenged to provide a more complete accounting of the costs and benefits of various development projects, as today's society expresses stronger attitudes toward environmental protection than in the late 1960s and 1970s (Holmes, 1979; Krutilla and Fisher, 1985; Palmer, 1986). The recreational benefits included in the economic analysis of water projects typically reflect only reservoir-based recreational opportunities created by impounds. Project analysts fail to include the values of stream-based recreation and associated environmental amenities that would be lost with the construction of river basin improvements. Their exclusion of stream-based benefits in traditional cost-benefit accounting is not surprising. Given the weak justification for many water projects, it is unlikely that agencies would include additional project costs voluntarily.

A more vocal public insistence on resource preservation has been accompanied by increasing sophistication in the ability to assign monetary values to resource benefits previously considered 'intangible'. The values used in cost-benefit accounting are measures of consumers' willingness to pay for the use and enjoyment of amenity or recreational resources. Three methods have been used to estimate willingness to pay: unit-day values, travel costs and contingent valuation (US Water Resources Council, 1983; Walsh, 1986). The unit-day value employs expert judgement to approximate the average willingness to pay for recreational use. Travel cost models use expenditures associated with trips to enjoy resources as a surrogate for willingness to pay. Contingent valuation sets up a hypothetical market and allows individuals to price environmental goods that do not customarily have dollar values assigned to them.

Using surveys to question individuals directly, the contingent valuation method can also separate total willingness to pay into 'recreational use' value and several 'preservation' values, including 'option' value (reserving the option to use the resource in the future), 'bequest' value (providing future generations with the ability to enjoy the resource), and 'existence' value (allowing the resource to exist for its own sake or for scientific purposes) (Sanders *et al.*, 1990). Scholarly debate has taken place, and undoubtedly will continue, over the merits and shortcomings of such economic methods (Willis, 1989). While the debate continues, these techniques for assigning economic values to resources for recreation or preservation offer policy makers the opportunity to include those resource values in cost-benefit analysis.

An Appalachian Example

Though new water project authorization has declined dramatically, some rural communities still look to the federal government for help in developing their struggling economies. The federal government, and most state governments, have been reluctant to take a lead role in comprehensive resource planning. Without a comprehensive view, they often lack important information, including data on the public's willingness to pay for stream recreation and preservation. Without a National Resources Planning Board to coordinate efforts, water resource agencies occasionally work at cross purposes. As a result, development projects usually follow the will of the influential group that amasses the largest constituency.

The example provided by one Appalachian state illustrates these points. People usually associate West Virginia with coal mining and the chemical industry, but the state has capitalized on growth in the travel and outdoor recreation industry. Abundant scenic resources and opportunities for outdoor recreation constitute the two most important reasons for travel in West Virginia (Goeke, 1986). Many natural features in West Virginia, and the recreational industries that have grown up around them, have turned the state into a popular playground for urbanites from across the eastern United States. Proximity to metropolitan areas, like Washington and Baltimore, guarantees a flow of out-of-state travellers that supplements the demand of in-state residents for use of the state's recreational resources.

Rivers and streams throughout the state provide a focus for much of the growth in outdoor recreation. The whitewater rafting industry attracts many visitors, especially to the New, Gauley and Cheat rivers, and their economic impact on surrounding communities is significant (Logar et al., 1984). Streams also support opportunities for trout fishing, kayaking and canoeing, and provide an important backdrop for other outdoor activities like hiking, biking, camping and hunting. These stream-based recreational activities can be enjoyed for a longer season than skiing, the other major recreational activity that generates significant income for the state. They hold great potential for stimulating even more local economic development when effectively marketed.

The Cheat River basin

The Cheat River basin, in north-central West Virginia, is one of many in Appalachia beset by the problems of declining coal, timber and manufacturing economies. Its rugged terrain offers numerous possibilities for recreation and tourism development, but the environment has suffered at the hands of careless resource extractors. Different constituencies are developing in response to two federal water management options: the construction of a multipurpose reservoir, or the protection of streams as part of the National Wild and Scenic Rivers Systems (Fig. 16.1). The need for an integrated review of the issues involved in economic growth and protection of the region's remaining resources has become pressing.

Some regional leaders favour an Army Corps of Engineers reservoir. They believe the structure will provide flood protection from the area's steep gradient streams and a much-needed boost to the local economy, both in construction industries and in reservoir-based recreation industries. Congress authorized, but never constructed, a dam on the Cheat River near Rowlesburg in the 1960s. Extensive flooding in West Virginia's Cheat, Potomac and Greenbrier river basins, in the autumn of 1985, renewed interest in flood control structures. In addition to a mainstem dam on the Cheat near Rowlesburg, a series of five dams on tributaries has been proposed to provide flood protection for Parsons and other communities downstream. The Corps of Engineers currently favours a reconnaissance study for flood control in the Cheat basin, but the likelihood of major dams is almost non-existent. New projects face state cost-share requirements and other regulatory hurdles (Palmer, 1986; Echeverria *et al.*, 1989), yet the recent completion of Stonewall Jackson dam and lake in central West Virginia has left area residents hopeful.

An alternate water resource management strategy would lead to Congressional designation of several Cheat basin tributaries under the federal *Wild and Scenic Rivers Act* of 1968, which protects outstanding free-flowing rivers in the United States from dams and other federal water projects (Coyle, 1988). The President's Commission on Outdoor Recreation (1987) recommended the extension of state and federal protection to additional river segments across the country. The US Forest Service responded with a review of streams and found seven in the Cheat basin eligible for further protection. Riparian private property owners along protected streams continue to hold title to their land and cooperate with the federal government in designing and implementing a management plan for the river corridor.

A mainstem dam at Rowlesburg would not necessarily be affected by designation of wild or scenic streams in the headwaters, but the tributary dams would. Perhaps more importantly, the two proposals reflect essentially different mind-sets about how to address the weak economy, resource management and flood control. Some see reservoir-based recreation as more profitable for the area. Others believe that the protection of outstanding natural resources is critical to the continued growth of the tourism and recreation industry in the state. Neither the Corps nor the Forest Service has attempted to document the economic value of the basin's existing streams. Communities also have been slow to explore non-structural adjustments to flooding, such as floodplain zoning, early warning systems and federal flood insurance (Bromley and Barrows, 1974; Holmes,

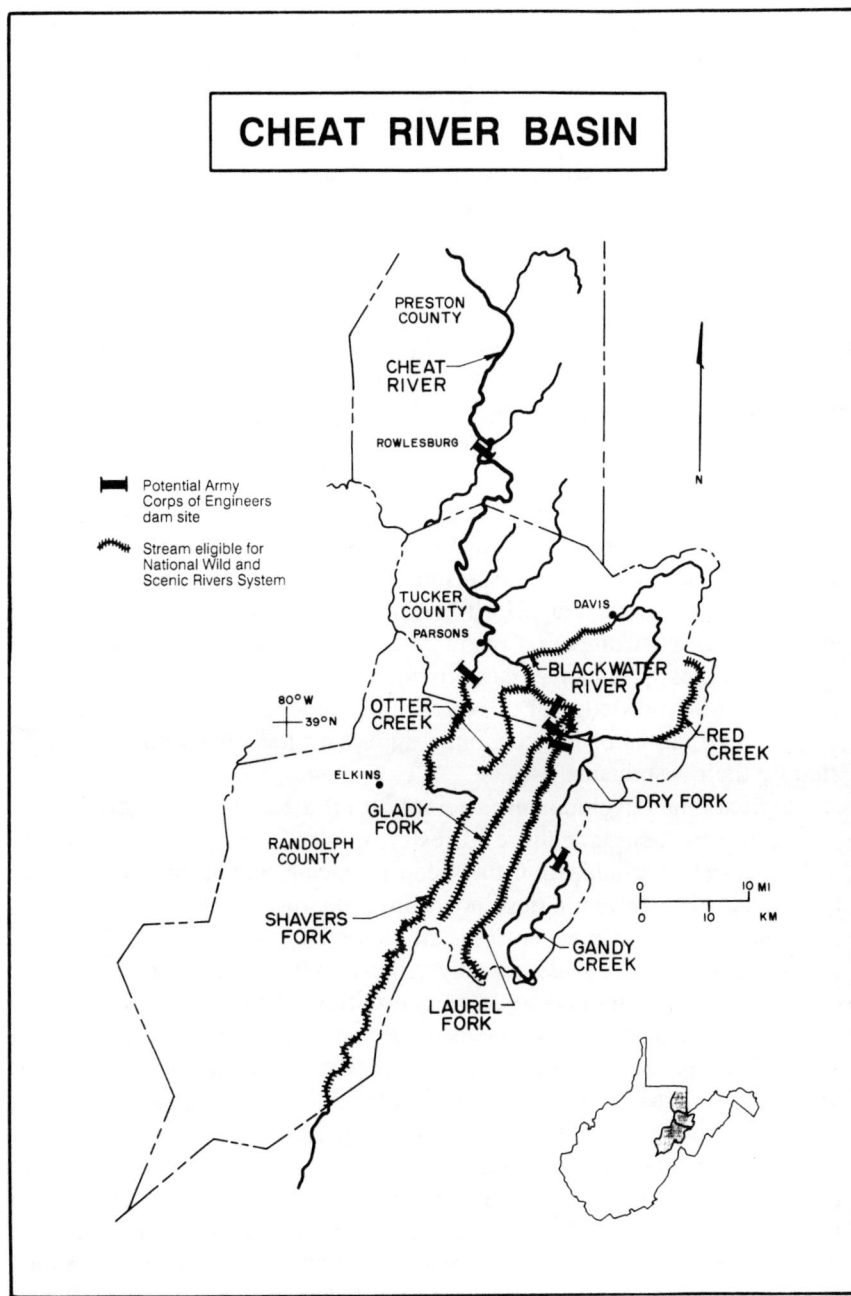

Fig. 16.1. Cheat River basin.

1979). Not surprisingly, all the parties who plan for and use resources in the Cheat River basin have failed to discuss the region's future at a common table.

Managing Water Resources for Sustainability

Political deals that led to federally funded water projects earlier in this century have virtually halted with the rise of environmentalism and a budget that can no longer support such heavily subsidized developments. Incomplete or questionable accounting resulted in the approval of many projects that dramatically altered river basins across the United States. Their promises of regional economic development largely went unfulfilled.

The changing global economy and tight federal budget require rural communities to restructure their economies, without relying on capital-intensive federal programmes like the water projects of previous decades. As restructuring takes place, local leaders need to recognize the import-ance of recreation and amenity resources that hold a key to employment diversification. Realigned priorities reflect more broadly based societal interest in all environmental resources, and the recognition that streams and other amenity resources can provide a more sustainable base for local employment. Knowledge of these resource values serves as a springboard for further analysis of the economic impact on local communities gener-ated by their sustainable uses.

Recreational development in remote rural areas is more likely to pro-vide new jobs than agriculture, forestry, mining or manufacturing activi-ties, in which mechanization has been replacing human labour. Neither individual rural communities, nor entire regions or states, can expect recreation and tourism alone to sustain them. But associated service-based employment does provide the diversification needed for healthy local economies. Communities should recognize the need to maintain the integ-rity of the environment, because its nature and quality often help to attract other types of economic activity. Areas that have not sacrificed their amenity resources are more likely to attract industries which are evaluating alternative sites to meet their raw material and market accessibility needs.

Environmental integrity, quality of life and economic diversification goals can only be met when residents and local leaders demand more of themselves through a 'bottom-up' campaign. Environmental protection programmes will continue to rely on federal and state funding, just as they have in the past, but rural residents must demand that elected officials be more responsive to the concerns of local regions. Federal and state programmes to aid small businesses will also be necessary to offer locally initiated economic growth and diversification the chance to flourish.

Coordination of all government programmes, whether environmental or economic, must become a priority.

Nearly 50 years have passed since a National Resources Planning Board convened. The US Congress needs to resurrect such a body to orchestrate the currently fragmented control that individual federal agencies exercise over the country's water resources. States should follow suit and provide better coordination and communication between their resource management departments. As public interest in environmental issues reaches new heights, government reorganization of resource management is likely to receive popular support. Restructuring in both rural employment and resource management are essential partners as communities reach beyond simple economic growth and strive for sustainable development.

Acknowledgement

This research was supported in part with funds provided by the US Geological Survey, Department of the Interior, as authorized by the *Water Resources Research Act* of 1984.

References

Armstrong-Cummings, K. (1986) Strategies for a sustainable future in Appalachia. In: Appalachian Center, *Proceedings of the 1986 Conference on Appalachia*. The Center, University of Kentucky, Lexington, pp. 9–15.

Berkman, R.L. and Viscusi, W.K. (1973) *Damming the West*. Grossman Publishers, New York.

Bromley, D.W. and Barrows, R.L. (1974) The changing nature of water resource investments: implications for community development. In: Field, D.R., Barron, J.C. and Long, B.F. (eds) *Water and Community Development*. Ann Arbor Science Publishers Inc., Ann Arbor, pp. 81–105.

Butcher, W.R. (1974) The role of water resources in community development. In: Field, D.R., Barron, J.C. and Long, B.F. (eds) *Water and Community Development*. Ann Arbor Science Publishers, Inc., Ann Arbor, pp. 59–79.

Chandler, W.U. (1984) *The Myth of the TVA: Conservation and Development in the Tennessee Valley, 1933–1983*. Ballinger Publishing Company, Cambridge, MA.

Clawson, M. (1981) *New Deal Planning: The National Resources Planning Board*. Johns Hopkins University Press, Baltimore.

Coyle, K.J. (1988) *The American Rivers Guide to Wild and Scenic River Designation*. American Rivers Inc., Washington DC.

Echeverria, J.D., Barrow, P. and Roos-Collins, R. (1989) *Rivers at Risk: The Concerned Citizen's Guide to Hydropower*. Island Press, Washington DC.

Goeke, P.E. (1986) *Travel in West Virginia: Re-evaluation of Growth in the 1980s.* Report for the Governor's Office of Community and Industrial Development, Bureau of Business Research, West Virginia University, Morgantown.

Haveman, R.H. and Krutilla, J.V. (1968) *Unemployment, Idle Capacity, and the Evaluation of Public Expenditures: National and Regional Analyses.* Johns Hopkins Press, Baltimore.

Holmes, B.H. (1972) *A History of Federal Water Resource Programs, 1800–1960.* USDA Miscellaneous Publication No. 1233, Government Printing Office, Washington DC.

Holmes, B.H. (1979) *History of Federal Water Resource Programs and Policies, 1961–1970.* USDA Misc. Publ. No. 1379, Government Printing Office, Washington DC.

Krutilla, J.V. and Eckstein, O. (1958) *Multiple Purpose River Development.* Johns Hopkins Press, Baltimore.

Krutilla, J.V. and Fisher, A.C. (1985) *The Economics of Natural Environments: Studies in the Valuation of Commodity and Amenity Resources.* 2nd edn. Resources for the Future, Washington DC.

Logar, C., Goeke, P.E., Rose, A.Z. and Davis, G.E. (1984) *Economic Impacts of Whitewater Boating on the Gauley River.* Report to the Army Corps of Engineers, Bureau of Business Research, West Virginia University, Morgantown.

Maass, A. (1951) *Muddy Waters: The Army Engineers and the Nation's Rivers.* Harvard University Press, Cambridge, MA.

Moore, M.R. (1991) The Bureau of Reclamation's new mandate for irrigation water conservation: purposes and policy alternatives. *Water Resources Research* 27, 145–55.

Palmer, T. (1986) *Endangered Rivers and the Conservation Movement.* University of California Press, Berkeley.

President's Commission on Outdoor Recreation. (1987) *Americans Outdoors: Report of the President's Commission.* Island Press, Washington DC.

Reisner, M. (1986) *Cadillac Desert: The American West and Its Disappearing Water.* Penguin Books, New York.

Sanders, L.D., Walsh, R.G. and Loomis, J.D. (1990) Toward empirical estimation of the total value of protecting rivers. *Water Resources Research* 26, 1345–57.

Smit, B. and Brklacich, M. (1989) Sustainable development and the analysis of rural systems. *Journal of Rural Studies* 5, 405–14.

Tolley, G.S. (1966) The impact of water investments in depressed areas. In: Kneese, A.V. and Smith, S.C. (eds) *Water Research.* Johns Hopkins Press, Baltimore, pp. 457–70.

US Water Resources Council (1983) *Economic and Environmental Principles and Guidelines for Water and Related Land Resources Implementation Studies.* Government Printing Office, Washington DC.

Wahl, R.W. (1989) *Markets for Federal Water: Subsidies, Property Rights, and the Bureau of Reclamation.* Resources for the Future, Washington DC.

Walsh, R.G. (1986) *Recreation Economic Decisions: Comparing Benefits and Costs.* Venture Publishing Inc., State College, Pennsylvania.

Warne, W.E. (1973) *The Bureau of Reclamation.* Praeger Publishers, New York.

Willis, K.G. (1989) Option value and non-user benefits of wildlife conservation. *Journal of Rural Studies* 5, 245–56.

17

LOCAL ECONOMIC INITIATIVES IN RURAL AREAS OF THE UK

Stephen Martin

Since the mid–1970s there has been a considerable growth in the number of local economic initiatives that have been undertaken in the UK. This has been partly due to continuing high levels of unemployment, and to the withdrawal of central government support for regional assistance. Work such as that of Fothergill and Gudgin (1982), which suggested that economic decline was the result of local supply problems, also provided a further boost to local economic intervention. One of the most widely adopted local economic initiatives has been the provision by the public sector of business premises. This chapter examines the impacts that this sort of local economic intervention has on employment levels and labour markets in rural areas.

Premises Provision in Rural Areas

The provision of premises has been supported by all the main political parties in the UK (Slowe, 1981) and has been welcomed by all tiers of government. Local authorities have seen it as a means of broadening the economic bases of their areas. Central government, though wary of involvement in a 'zero sum' game, which merely shifts jobs from one area to another, has supported it as a means of promoting the growth of small firms and encouraging new firm formation. However, rural local authorities have generally been less interventionist than their metropolitan counterparts. As a result, most of the publicly funded workspace in rural England has been financed, at least in part, by the Rural Development Commission (the development agency for rural England) which has provided more than 2400 workshops comprising in excess of 3.5 million square feet.

The rationale for the earliest workshops (provided in the late 1950s) was founded firmly in 'growth pole' theory. They were intended to attract in-moving firms to locate in 'Trigger Areas' thereby boosting the local economy, catalysing increased economic activity and reversing the cumulative downward spiral of disinvestment (Tricker and Martin, 1984). By the 1970s, however, they were seen primarily as a means of arresting rural depopulation and provision was concentrated in new priority areas known as 'Special Investment Areas'. By the early 1980s it was clear that population levels were in fact increasing in most of these areas. The workshop building programme was, therefore, re-packaged and presented, not as the solution to a demographic problem, but as a means of increasing the number and range of employment opportunities in areas that had traditionally been over-dependent on one or two major industries, such as agriculture and mineral extraction (Rural Development Commission, 1984).

Within 5 years the problems of demographic imbalance in rural communities associated with the counter-urbanization of the elderly and middle classes meant that the creation of new jobs, through workshop provision, was being seen as a means of retaining young local people in rural areas. However, by this time the main emphasis was upon facilitating the formation and expansion of new enterprises, which were thought to be vital to macro-economic recovery (Rural Development Commission, 1988), and intervention of any sort was increasingly difficult to justify to sceptical Treasury minister and officials. Public sector provision of workspace in rural areas was, therefore, being promoted primarily as a way of correcting market failure in the industrial property market.

The Impacts of Premises Provision

Previous studies

Over the last 10 years several studies have evaluated the impact of premises provision in rural areas in the UK. Different researchers have adopted different approaches to measuring employment impacts and, as a result, it has been difficult to compare their findings. However, they have all demonstrated that premises provision has had significant beneficial impacts on the local economies of rural areas. Four separate studies of the effects of premises provision in different parts of England and Wales (Tricker *et al.*, 1983; Whitby and Willis, 1983; Thomas and Drudy, 1987; Willis and Saunders, 1988) have demonstrated that the provision of premises has created very few new jobs in terms of the national economy. But between 80 and 90% of the jobs provided by businesses that have occupied publicly funded workshops would not have existed in the same local areas if the premises had not been built.

However, each of these studies was a 'cross-sectional' analysis at just one point in time; as a result, very little is known about the medium- and long-term effects of premises provision. In addition, because attention was focused on the numbers rather than the types of jobs provided, there is very little knowledge of the wider labour market impacts. It was in order to investigate these impacts, and the longer-term employment effects of workshop provision, that a longitudinal study of the impacts of premises provision in the English Midlands was undertaken.

Case study areas

The case study areas, though all in the same region, included a range of settlements of different sizes located at varying distances from the nearest metropolitan areas. Each area was relatively small, with a radius of less than 10 miles incorporating one settlement and the surrounding 'area of pull' (the 'target area' defined by the Commission). Three of the settlements were market towns with large agricultural hinterlands and populations ranging from 4000 to 7000 people. The other five settlements were much smaller, with populations of between 350 and 1300. One town and two of the villages were situated close to two major motorways and were just 12 miles from a large urban area. A second town and one of the villages were situated in the Peak District National Park, which is more remote, but within commuting distance of three major conurbations. The third town and the two remaining villages were in more remote locations in excess of an hour's drive from the nearest city (Fig. 17.1).

By 1984, 56 workshop units had been built in the eight areas – 36 in the market towns and 20 in the villages: 44 (78%) were occupied. The occupant businesses were involved in a wide range of activities. Two-thirds were engaged in some form of manufacturing. Most were relatively new. One-third had been started up in the workshops; 55% had been founded in the last 5 years; 80% were less than 10 years old. All but six of the businesses that had moved into the workshops from other premises had previously been located less than 20 miles away.

The managing directors or owners (usually the same person) of the businesses were interviewed in 1984, 1986 and 1991, and these three surveys provide the basis for longitudinal assessment of the impacts of workshop provision over a period of 7 years.

Employment generation

Short-term impacts

In 1984, the businesses occupying the workshops in the case study locations employed 360 staff, of which a third were part time. By aggregating the hours worked by the part-time employees, and dividing this by a notional

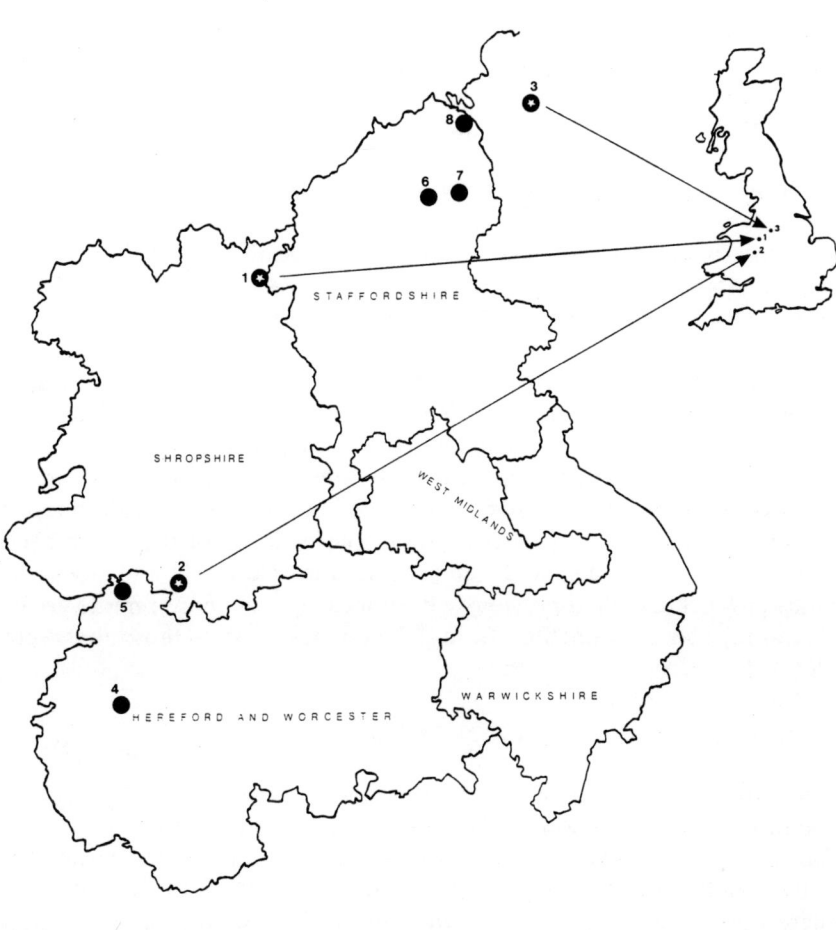

Fig. 17.1. Location of case study areas. 1 – Market Drayton; 2 – Ludlow; 3 – Bakewell; 4 – Weobley; 5 – Leintwardine; 6 – Ipstones; 7 – Waterhouses; 8 – Longnor; ✪ – Market town; ● – Associated village.

average working week of 40 hours, it was calculated that the businesses provided a total of 319.5 full-time equivalent (FTE) jobs. As might be expected, given that nearly half the businesses were new start ups, employment levels in the businesses had increased rapidly since moving into the workshops. However, most of the managing directors/owners believed that if they had not been able to move into the workshops, they would have found alternative premises and their staffing levels would have been the same. As a result, in their judgement, only 66 FTE jobs had been created by the provision of the workshops.

The local employment impacts of the provision of the workshops were, however, much greater. Many of the occupant businesses would have located outside the case study areas if they had been unable to move into the workshops. Analysis of their alternative locations, and the employment levels in the 40 businesses, suggested that 88% of the FTE jobs they provided would not have existed in the case study areas if the businesses had been unable to move into the workshops. These findings are remarkably consistent with the results of previous studies of the employment effects of advance workshop building (described above), and not surprising given that the target areas defined by the Commission are small.

Comparisons between the case study areas revealed that the largest numbers of new *local* jobs had been provided by non-indigenous businesses which had relocated from urban areas. The majority of these had been attracted to those case study areas that were closest to the towns or cities in which they had previously been located. The more remote areas, therefore, had not benefited from as many new jobs as those that were closest to metropolitan areas.

Longer-term impacts

The longer-term effects of workshop provision were assessed by tracking the businesses that had occupied workshops in the case study areas over the next 7 years. The 1986 survey showed that many of the businesses had expanded very rapidly in the 2 years between 1984 and 1986. Although there were wide variations in the rates of increase between areas, overall employment levels had increased by 61%. This suggests that, in the longer term, the scale of employment generation associated with premises provision may be much greater than has been suggested by previous studies which have provided only a 'cross-sectional' view.

The managing directors/owners of most of the businesses expected them to continue to grow over the following years. Overall they anticipated a 44% increase in employment between 1986 and 1988. This represents a slower rate of growth than had occurred in the previous 2 years, but a total increase of 132% between 1984 and 1988 (Fig. 17.2). However, much of this additional employment growth seemed likely to occur outside

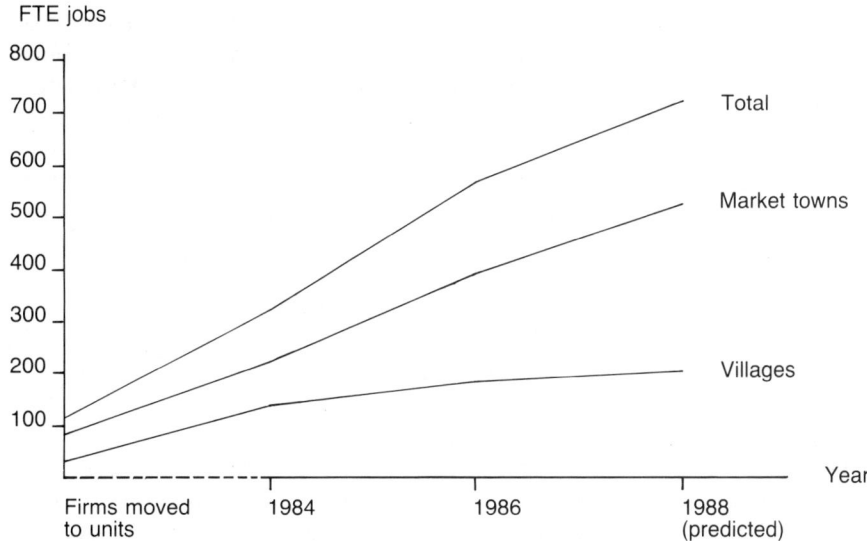

Fig. 17.2. Changes in the number of jobs provided in workshops.

the case study areas. Two businesses had already moved away and many more had outgrown their original workshop premises. Very few of the managing directors/owners believed that there was any chance of finding larger premises in the case study areas and many believed they would therefore be forced to relocate. Four expected to move their entire operations out of the case study areas in the near future and another 12 expected to open branch plants elsewhere.

In addition to the likely exodus of jobs from the case study areas, several of the managing directors/owners who expected to remain in the case study areas believed that, because of the shortage of larger premises, their future growth would be severely restricted – thus the lower anticipated rate of employment increase between 1986 and 1988, compared with 1984 and 1986 (Fig. 17.2).

Preliminary analysis of the 1991 survey suggests that some of the fears expressed in 1986 were unfounded. Less than a quarter of the businesses had in fact moved their entire operations out of the case study areas, and it is clear that the provision of premises does, therefore, produce sustained benefits for the local areas in which they are located. However, the rate of employment growth in the businesses that remained in the case study areas was much slower between 1986 and 1991, than it had been between 1984 and 1986. This may have been the result of a number of factors, including macroeconomic conditions, but in many cases it was caused by the shortage of large premises in the case study areas.

The 1991 survey also suggests that non-indigenous businesses (existing firms that moved into the case study areas because of the availability of workshops) have a greater propensity to move away in later years. Since they expanded more rapidly than local firms or start-ups, and have therefore created the largest numbers of new local employment opportunities, their relocation implies a considerable loss of jobs from the case study areas. This may vindicate the emphasis that has been given in recent years to encouraging the growth of existing local businesses and new start-ups, rather than attracting in-moving firms.

Local labour market impacts

Short-term impacts

BROADENING THE LOCAL ECONOMIC BASE
In addition to creating new jobs, one of the objectives of workshop provision was to help to maintain 'balanced communities' in rural areas by increasing the range of jobs available locally, especially for young people. Analysis of the types of jobs provided by the businesses in the workshops suggests that they have provided job opportunities traditionally lacking in rural areas. In 1984, 40% of the jobs they provided were skilled and less than a third were unskilled. An analysis of a sample of matched firms of similar sizes and sectors, located in or near to the case study areas, showed that the latter provided significantly fewer skilled jobs. The employees' perceptions support this view. Only 26% of those who had been in employment prior to taking up a job in one of the businesses in the workshops considered that their previous job had been skilled; 41% considered it to have been unskilled. By comparison, 40% considered their job in the workshops to be skilled and just 29% reported that they were engaged in unskilled occupations. Two-thirds considered that their present job (in the workshops) offered improved working conditions, increased pay, better prospects, or more interesting work.

The businesses in the workshops had also provided more jobs for young people than the matched firms. Nearly 66% of the jobs in the workshops had been taken by people between the ages of 25 and 45 years, and 29% had been filled by people under 25 years. Since many lived locally and had young families, this is likely to have had a positive demographic effect in the case study areas.

FREEING UP NEW JOBS
The second-tier impacts of premises provision (i.e. those resulting from the freeing up of jobs in other businesses) depend on the previous employment status of the employees who had taken new jobs in the workshops (Martin and Tricker, 1989). There are three possible scenarios (Fig. 17.3). First, they may have previously been registered as unemployed, in which

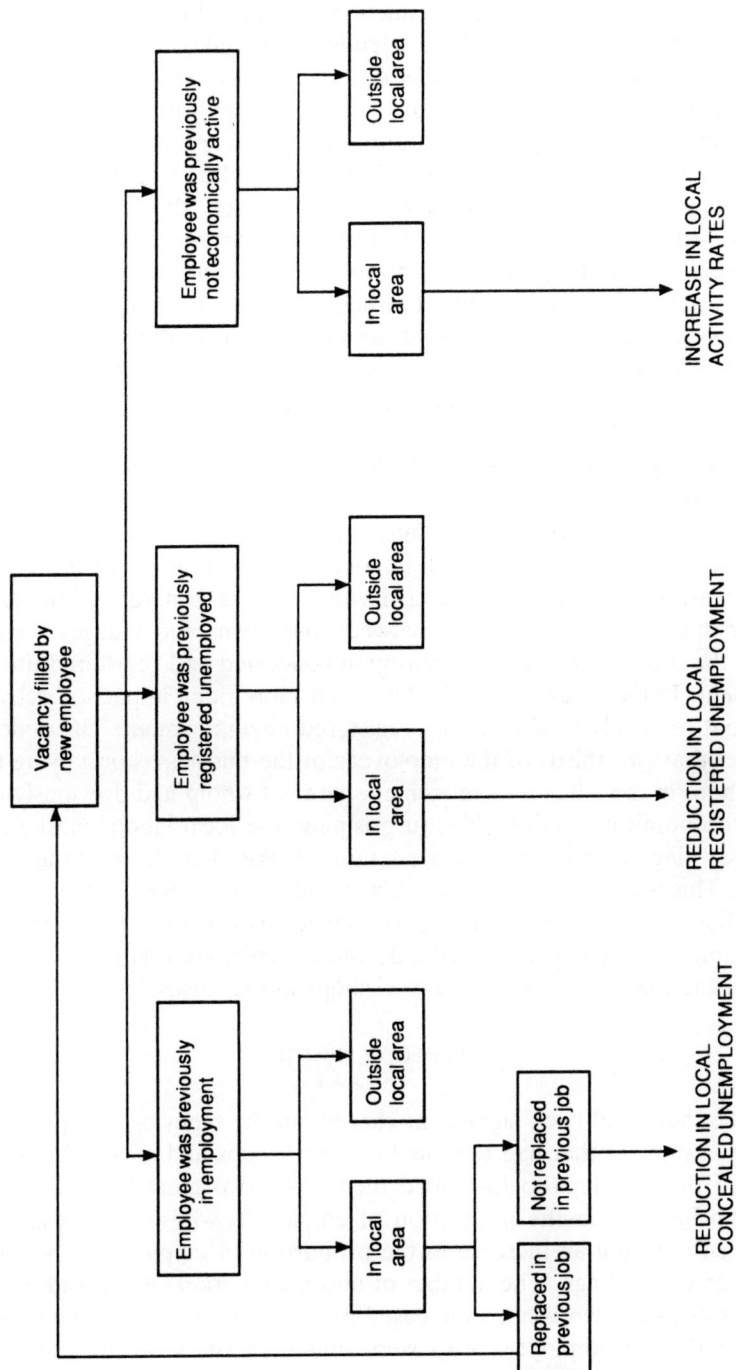

Fig. 17.3. Model of the second-tier impacts on local labour markets (Source: Martin, 1988).

case the result will have been a reduction in the level of registered unemployment. Second, they may have been without paid employment but not registered as unemployed. In this case the provision of new jobs will have increased activity rates. Third, they may have previously been in employment, in which case the nature of the impact on the labour market will have depended upon whether they were replaced in their previous job and, if so, by whom. If they were not replaced, the result will have been a reduction in 'concealed' unemployment or underemployment. If they were replaced, a second iteration of employment effects will have been initiated, the end result of which will have been determined by the previous employment status of the person who replaced them. Each sequence of job changes will have continued until it was terminated by a reduction in registered or concealed unemployment or an increase in activity rates.

By tracing the sequences of job changes associated with the provision of new employment, it was found that by 1984 the provision of the workshops had led, within the case study areas, to a reduction in registered unemployment of 57, a reduction in concealed unemployment of 26, and an increase in activity rates of 56 (Martin, 1988). The nature of the local labour market impacts differed between the towns and villages. In the towns the major impacts had been upon concealed and registered unemployment. In the villages there had been far more of an impact on activity rates but relatively little effect on registered unemployment. This reflects the fact that two-thirds of the employees of the businesses occupying the workshops in the villages were women, many of whom had previously not been 'economically active'. Not surprisingly, the local labour markets of the five village case study areas were more 'leaky' than those of the three towns. This was probably due, in part, to labour shortages in the villages – the businesses in workshops in the villages had recruited only 58% of the employees from within a 10-mile radius, compared with 77% in the case of the businesses located in workshops in the towns.

Longer-term impacts

By 1986 there had been significant changes in the types of jobs provided by the businesses that had occupied the workshops and in the characteristics of the people who had filled them. The two most marked trends were an increase in the proportion of employees who were engaged in unskilled jobs and an increase in the proportion of employees who were under 25 years of age. The number of unskilled workers employed by the businesses in the workshops increased by 137% between 1984 and 1986. The number of employees who were under 25 years old increased by 140%. By 1986, those in unskilled jobs comprised 34% of the workforces of the businesses, compared with 29% in 1984, and the proportion of the employees who were under 25 years had increased from 30 to almost 50%.

The changes in the composition of the workforces demonstrate that the impacts of workshop provision on the local labour markets may change over time and highlight the need for longitudinal studies. The precise causes and consequences of the changing recruitment patterns are a matter for further research. However, the 1986 survey suggests that the changes in both the age and skill profiles of employees were largely due to the expansion of the occupant businesses and the resulting increase in their need for operatives to undertake routine unskilled tasks.

Conclusions

Whether workshop provision in rural areas is judged to have been a success or not depends on which of its objectives is used as the measure. Like previous research, the work described in this chapter demonstrates that premises provision has helped to increase the number and range of job opportunities in target areas. It seems that, in the medium term, the number of jobs provided is likely to continue to increase, albeit at a slower rate, and that, as they expand, businesses occupying workshops recruit increasing numbers of young people and unskilled workers – two of the main groups which it is important to retain in rural communities if they are not to become dominated by elderly people and middle-class commuters.

At the local level, the provision of new job opportunities has helped to alleviate the three main employment problems facing rural areas, namely: unemployment, underemployment and low activity rates, especially among women. Contrary to the expectations of the policy makers responsible for the provision of publicly funded premises in rural areas (who expected it to reduce registered unemployment) one of the main effects of workshop provision in the early 1980s was to 'soak up' concealed unemployment. This may have been because concealed unemployment is more widespread in rural labour markets than had previously been realized. Alternatively, it may have been the result of the severe economic recession that occurred at this time. Too little is known about the detailed workings of rural labour markets for definite conclusions to be reached at this stage. However, the present study has demonstrated that sequences of job changes associated with the creation of new jobs are often much shorter and, therefore, easier to trace than had been assumed in the past. Hopefully this will stimulate further research into the workings of the rural labour market.

It is also clear that many of the advantageous employment impacts of premises provision leak out of small target areas, such as those defined by the Rural Development Commission. This raises questions about the usefulness of targeting initiatives on such small areas, especially when their boundaries have been determined on an arbitrary basis, or drawn so as to coincide with local authority areas which are much smaller than local travel-to-work areas. The mobility of most employees suggests that concentrating workshop provision in key settlements may be more cost-effective than building them in small villages. However, this would undoubtedly have an adverse effect on employment prospects for those without the means to travel to work outside their villages.

Finally, the existence of 'employment ceilings' in existing premises, the lack of alternative larger premises in rural areas, and the fact that non-indigenous firms – which create the largest numbers of jobs – have the greatest propensity to relocate to metropolitan areas, suggest that there is a continuing role for the public sector in promoting the supply of suitable business premises, particularly at the upper end of the 'workspace ladder'.

References

Fothergill, S. and Gudgin, G. (1982) *Unequal Growth: Urban and Regional Employment Change in the UK*. Heinemann Books, London.

Hodge, I.D. and Whitby, M.C. (1979) *New Jobs in the Eastern Borders: An Economic Evaluation of the Development Commission Factory Building Programme*. Agricultural Adjustment Unit, University of Newcastle upon Tyne, Newcastle.

Martin, S.J. (1988) Evaluating local economic initiatives: an assessment of the Rural Development Commission's advance factory building programme. Unpublished PhD thesis, Aston University, Birmingham.

Martin, S.J. and Tricker, M.J. (1989) The impact of rural employment initiatives on local labour markets. *Nederlandsche Geografische Studies* 92, 176–84.

Rural Development Commission (1984) *The First Seventy-five Years*. RDC, London.

Rural Development Commission (1988) *Promoting Jobs and Communities in Rural Areas*. Annual Report 1987–1988, RDC, London.

Slowe, P. (1981) *The Advance Factory in Regional Development*. Gower, Aldershot.

Thomas, I.C. and Drudy, P.J. (1987) The impact of factory development on growth town employment in Mid-Wales. *Urban Studies* 24, 361–78.

Tricker, M.J., Bozeat, N. and Martin, S.J. (1983) *An Evaluation of the Development Commission's Activities in Selected Areas of England*. JURUE, Birmingham.

Tricker, M.J. and Martin. S.J. (1984) The developing role of the Commission. *Regional Studies* 18, 507–14.

Whitby, M.C. and Willis, K.G. (1983) New jobs in Mid-Wales. Unpublished study for the Development Board for Rural Wales, DBRW, Newtown.

Willis, K.G. and Saunders, C.M. (1988) The impact of a development agency on employment: resurrection discounted? *Applied Economics* 20, 81–96.

18

COMMUNITY DEVELOPMENT AND CHANGING RURAL EMPLOYMENT IN CANADA

Christopher Bryant

Canada has experienced significant sectoral shifts in employment since the mid-20th century as the economy has moved rapidly towards a postindustrial economy (Bell, 1973). Other structural changes, less well-broadcast, are also involved, for example the reorganization of economic activities in terms of enterprise size, labour-management relationships and the role of capital. These shifts are responses to a complex pattern of forces involving technological change, new consumer needs, increasing interdependency of the global economic system and the emergence of significant trading blocks.

How have these changes affected rural areas? Has the response of rural areas been reactionary or more proactive? What are the prospects for the last decade of the 20th century? These are the main questions addressed in this chapter, the main theme of which is the relationship between local level responses and rural employment change.

First, a conceptual framework is presented that links the major forces affecting economic activities, their geographic scale of operation and the respective roles of communities and upper levels of government. Then, the nature of local development is discussed and the relationships between local and community development processes teased out.

A Conceptual Framework

The literature on community (economic) development is considerable, both in Canada and elsewhere. Much of it is characterized by case study approaches cast frequently in terms of 'success' or 'failure' stories, although some recent attempts have been made to synthesize the Canadian literature and experience (see Bryant and Preston, 1987a, b; Douglas, 1989; Dykeman, 1989, 1990; Perry, 1989; ICURR, 1990; O'Neill, 1990).

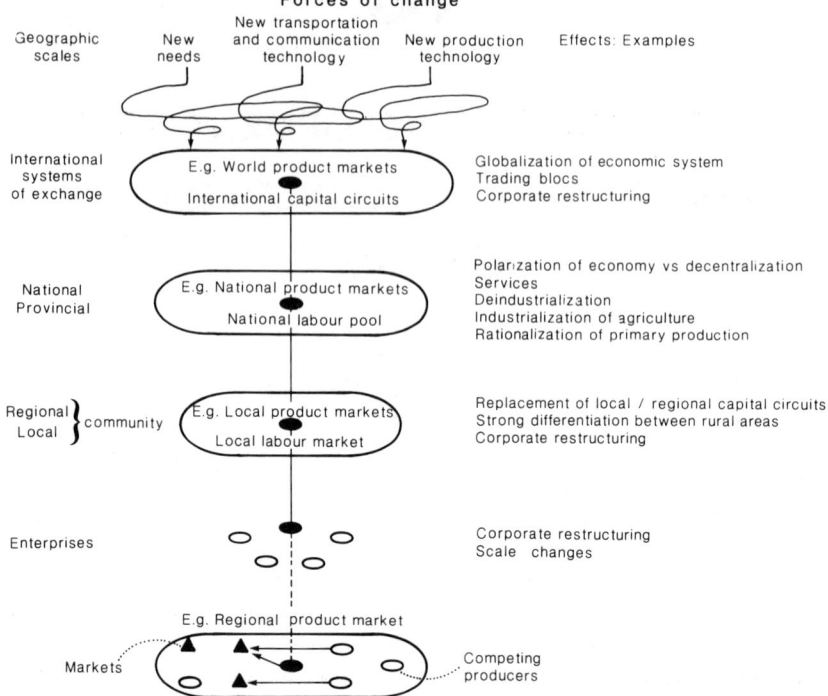

Fig. 18.1. Systems of exchange and economic activities.

How can links be established between the broad changes occurring in society, employment change in rural areas and the role of the community in these processes? The framework proposed has three cornerstones: systems of interactions; a geographic differentiation of 'rural'; and the link with community development processes.

First, economic activities function within different systems of exchange (for example, product/service markets, capital markets) operating at different geographic scales (Fig. 18.1) (Bryant, 1988; Bryant and Coppack, 1991; Bryant and Johnston, 1991). Each is composed of a set of nodes (for example, for the produce marketing system these would include the production unit and its competitors, retail and wholesaling operations and processing industries), and interactions between the nodes (for example, flows of produce) using particular transportation and communication media along particular networks.

Changes in any component of a system of exchange can have an impact on the production units. Some changes can be linked to the broad forces underlying the transformation of the social and economic system. In the context of postindustrial society, three broad sets of forces are important: (i) the emergence of 'new needs' as life-styles and values have changed; (ii) the development of new transportation and communications technology creating an increasingly open economic system, despite the continued existence of many barriers to trade, and (iii) the development of new technology affecting production processes, sometimes favouring large-scale and sometimes small-scale production processes (Bryant, 1988). All have influenced the restructuring of rural employment, as well as many other facets of the organization of rural areas.

Second, 'rural' space is not homogeneous. A significant distinction can be made between rural areas in metropolitan and those in non-metropolitan areas. This distinction is related to differences in market access, despite modern communication and transportation technology, as well as differences in access to different labour markets. Other differences include the nature of the economic base. These are partly related to the metropolitan and non-metropolitan distinction and partly to different types of 'rural' activities such as farming, forestry and fishing, tying communities into different although often overlapping systems of exchange.

Third, collective intervention is one of the sets of forces that potentially modifies systems of exchange. From the 1960s to the early 1980s, 'state' intervention (federal and provincial in Canada) developed either to reduce the negative impacts of restructuring for the communities and regions concerned, or to lead the way in developing new directions. Because of difficulties encountered, more and more attention has focused on the role of communities. This local and community action represents an attempt to respond to the forces of change either by reacting, adapting and/or by developing more proactive positions to improve the social and economic well-being of the community. These actions may involve attempting to re-position a community within existing systems of exchange (for instance, through marketing or encouraging the development of new enterprises), or even to create new systems of exchange (for example, through the development of new tourist attractions).

Perspectives on the Impacts of Change

The impacts of change can be seen from three perspectives. First, there are impacts at the broad macro- or meso-scale in terms of the degree of concentration of population and employment opportunities in urban and more generally metropolitan areas. The 'population turnaround', much touted in the United States in the 1970s and early 1980s, was never a particularly important phenomenon in Canada (Joseph *et al.*, 1988). Indeed, population concentration in Canada's major metropolitan regions has either remained stable or even increased (Bryant, 1991b). Second, there has been the extensively documented shift towards an economy dominated by service employment (see Bryant, 1991c; Coffey, 1987).

Third, there have been significant changes in the relationships between the different factors of production in the production processes in rural areas and in the relationships between producers and the market, for instance as reflected in changing patterns of firm size, with technology favouring continued increases in firm size in some sectors but greater stabilization and even decreases in others, and the globalization of both capital and markets, particularly in sectors characterized by large-scale production.

The impacts of the forces of change have been strongly differentiated within the Canadian space economy. One of the principal reasons is differentiation in the linkages between rural areas and the urban-industrial complex.

Rural areas in the urban field

The main urban fields extend from 150 km to 200 km around the major urban and metropolitan areas. They contain significant rural areas, including important portions of the nation's agricultural resource base and a significant share of other forms of employment (see Bryant, 1980). With the rise of industrial society and the subsequent development of post-industrial society, these rural areas have undergone complex changes in economic structure.

First, the industrial model of agricultural production has continued to develop there as elsewhere, fuelled by labour-reducing technological change and encouraged by easier access to a greater variety of labour markets and employment opportunities in the major urban fields. In those rural metropolitan areas most accessible to the urban cores, major community disruption has not accompanied the reduction of agricultural labour, technological change and farm consolidation; frequently, non-farm population growth has more than compensated for the reduction in the agriculturally supported population, stabilizing the retail and service sectors of many small towns there (Coppack, 1985).

Second, other stabilizing forces arise from the more diversified farm structure and the significant development of various types of part-time and hobby farming often observed in these areas. Third, some urban field settlements are able to draw urban consumers out into the countryside either to mix shopping and a 'rural' experience or simply as 'tourists'.

In the outer parts of urban fields, however, the industrial model of agricultural production has led to the decline in agricultural populations and employment, and there has been very little alternative development to prevent the stagnation and even decline of the community base.

The rural peripheries

Beyond the major urban fields, the many narrowly based rural economies, for instance agricultural areas with their supporting service centre network or the many single industry towns dependent upon resource exploitation, have been vulnerable to macro-scale forces such as globalization of markets. Rural peripheries function in increasingly national and international systems of exchange, and they have frequently had to 'roll with the punches' (Decter, 1989), without the stabilizing effects of the more diversified employment opportunities that are often found in rural zones accessible to metropolitan areas.

Table 18.1. Share of the three main CMA[a] in their respective provincial populations, 1976–1986.

	% share of provincial population in each CMA			
	1976[b]	1981	1981[c]	1986
Montréal CMA	45.0	43.9	44.5	44.7
Toronto CMA	33.9	34.8	36.3	37.7
Vancouver CMA	47.3	46.2	46.2	47.9

[a]CMA: Census Metropolitan Area.
[b]Based on 1981 boundaries.
[c]Based on 1986 boundaries.
Source: Census of Canada.

In such rural peripheries, it is not surprising that many observers tend to think of community development as a marginal activity, a process that reacts to impacts lying outside the control of rural areas. The continued demographic, economic and political weight of the main metropolitan regions in Canada reinforces this perspective (Table 18.1), as does the fact that small communities across the country appear to be having difficulty holding their own (Table 18.2). Community development, however, needs to be seen from the perspective of the local community. In the economic development domain, successful local initiatives aimed at rural employment replacement and maintenance may not make a strong showing on the national map. But for a rural community, saving or adding a dozen jobs may be a major achievement, as might restricting a decline to 5% rather than 10%.

Local and Community Development

The early collective response to the issues facing rural peripheries and other rural areas without a rich web of exchanges and relationships with

Table 18.2. The distribution of Canada's population by size of municipality, 1981–1986.

Municipality size	Number of municips.		Population (millions)		% national population		% change population
	1981	1986	1981	1986[a]	1981	1986	1981–1986
>50000	75	78	11.7	12.4	48.1	49.3	6.6
5 – 49000	578	599	7.4	7.7	30.2	30.2	4.0
<5,000	5057	5332	5.3	5.2	21.7	20.5	−2.0
Total	5710	6009	24.3	25.3	100.0	100.0	4.0

[a]Excluding Indian communities and reserves for which enumeration was incomplete.

	1981[a]		1986[b]		% change
	millions	%	millions	%	population
Within CMAs/CAs[b]	18.3	75.2	19.2	75.8	5.1
Urbanized core	16.3	66.8	17.0	66.9	4.4
Suburban zone	0.4	1.6	0.4	1.7	9.3
Rural zone	1.7	6.8	1.8	7.2	10.9
Beyond CMAs/CAs	6.1	24.9	6.1	24.0	0.7
Urban	2.0	8.1	2.0	7.8	− 0.4
Rural	4.1	16.8	4.1	16.3	1.2
Total	24.3	100.0	25.4	100.0	4.2

[a]Including Indian communities and reserves for which enumeration was incomplete. Based on 1986 boundaries.
[b]CMA: Census Metropolitan Areas; CA: Census Agglomeration.
Source: Census of Canada.

the urban–industrial complex was essentially one driven by federal and provincial governments, the classic 'top-down' approach to regional disparities. The history of programme after programme, frequent bureaucratic reorganization, and millions of dollars of public funds is not a particularly edifying one (see Savoie, 1986). Suffice it to say that mounting dissatisfaction with the 'top-down' approach in Canada by the late 1970s (as elsewhere, see Stohr and Taylor, 1981) led to increasing attention being paid to local and community involvement in economic development. The questions now posed are: What is the nature of this local and community effort and what has been and could be its role in changing rural employment?

Local economic development implies an effort through local initiative using a significant dose of local resources to influence the processes of change in economic activities and to meet locally defined needs and objectives more effectively through locally selected strategies (Coffey and Polèse, 1985; Bryant and Preston, 1987a, c). Locally defined needs and objectives are presumably more sensitive to local conditions and circumstances; involving local initiative provides greater commitment; and using local resources provides greater leverage in developing partnership operations with outside interests. Local development implies there is a planning process, that local leadership exists to take some initiative, and that needs and objectives encompass both economic and non-economic matters.

In Canada, the local development effort varies enormously in terms of organization, philosophy, the enabling environment presented by upper levels of government and the mix of reactive versus proactive strategies. Clearly, a local approach has the potential to produce great variety. Such variety is in itself not a problem, except perhaps from a narrow bureaucratic perspective. Organizational forms reflect partly what already exists as well as particular mixes of personalities and many other factors. Strategies reflect the specificity of the existing economic base as well as potential opportunities.

Two contrasting paradigms

At the risk of oversimplification, two approaches at opposite ends of the spectrum of 'local' development can be identified. One is based on an industrial development model frequently drawing on development strategies and processes used in major urban and metropolitan areas, whereas the other is focused more on community involvement in the improvement of quality of life, economic development being clearly a means to an end.

Selected characteristics of each approach are given in Table 18.3. Under the 'industrial development' approach, community input tends to be more limited, the goals tend to be defined more narrowly, economic development strategies tend to be developed as in-house documents as opposed to having a broader community base, and the extent to which responsibility for economic development planning and implementation in the community is decentralized is limited (Bryant and Preston, 1990).

Until the 1970s, the industrial development approach focused on manufacturing activity as the main source of investment and development; only more recently has commercial and service development become a generally accepted part of local economic development. In the community development approach, economic, social and cultural activities are all part of the approach. The industrial development approach furthermore is characterized by a strong emphasis on external sources of investment and business development, while the community approach accords a much greater role to local sources both of investment (for example, developing local and regional capital circuits) and of business development and entrepreneurship generally.

Team-building approaches in the industrial development approach tend to be limited, frequently involving an elite group from the business community and local political circles, whereas the community development

Table 18.3. The industrial development and the community (economic) development paradigms: selected characteristics.

Selected characteristics	The industrial development paradigm	The community (economic) development paradigm
Community input and style of process	Community input limited Emphasis on jobs and local tax generation Economic strategies as in-house documents 'Control' vested in an 'elite' group	Broad-based community input Jobs and local taxes and economic development generally as a means to an end Economic strategies owned and shaped by the community Responsibility for organization and implementation decentralized
Sources of investment and development	Strong emphasis on manufacturing (but more recently including services and commerce) Strong emphasis on external sources of investment and business development	Broader emphasis including both economic and social activities Greater role accorded to endogenous business development, investment and entrepreneurship
Team-building	More limited	A critical component
Information analysis and transfer	Marketing aimed at external targets Limited direct counselling services to business	Greater attention to internal targets Proactive counselling of small business
Reduction of financial barriers	Transfer of information on incentives. Extent of local aid to business varies greatly between provinces due to differences in legal environment in which municipalities function	
Fostering an entrepreneurial climate	Less attention (at least in the early days)	A critical component Proactive work with local education institutions

approach would be characterized by a much greater representation of community stakeholders. In the area of information analysis and transfer, the industrial development approach is oriented to promotion and marketing to targets outside the community, with only limited counselling to the actual business operations. In the community development approach, much greater attention is placed on marketing to the community and its actual and potential entrepreneurs and investors (Bryant *et al.*, 1988). Small business development, therefore, naturally tends to play a very significant role in the community development approach.

In terms of reduction of financial barriers to development, both approaches incorporate the transfer of information on government grants and other forms of financial aid to business development. However, in terms of the creation of an entrepreneurial climate, the community development approach can be expected to give greater emphasis to the creation of an environment favourable to business and entrepreneurial development.

These are presented as two extremes. The industrial development approach was epitomized in many of the local and regional programmes aimed at economic development during the 1970s; it is frequently thought of as a city-based approach. The level of specialization implied in the profile (Table 18.3) in terms of sectoral orientation, focus on external marketing targets and on a relatively narrow base of involvement is partly explained by the specialization possible in large municipal corporations.

The community development approach often tends to be thought of as linked to rural areas and other small communities. However, there have been significant changes that have affected both approaches. First, local efforts of all shapes and sizes are increasingly being cast into a longer-term planning framework, the strategic planning framework, even though there are enormous differences in its organization and management (Table 18.3). Second, the quality of the information and analysis utilized in strategic economic development planning has improved greatly in the last 10 years.

The extreme form of the industrial development paradigm is breaking down in certain respects. It is increasingly recognized that the economic development effort is really a means to an end. Thus, it is increasingly more common to see the industrial development approach modified by greater involvement of a broader range of community stakeholders. Furthermore, while big city environments still favour specialization between, say, different departments in a municipal corporation, there is a greater recognition of the need to ensure coordination and cooperation between the different departments and groups involved in the overall community development process.

On the other hand, not all rural areas and small towns are characterized by the community development approach to economic development.

Indeed, many such areas adapted some of the characteristics of the indus-trial development paradigm, copying from the perceived successes of the larger cities. Many, therefore, became characterized by a strong focus on external marketing and the preparation of promotional materials, a strong emphasis on physical infrastructure as opposed to the softer infrastructure of education and social development, and a preoccupation with the 'movers and shakers' in a community. These characteristics combined in many situations to create a relatively narrow and almost local brand of the 'top-down' approach to economic development!

The proactive activity associated with this approach tends to be rather narrow in terms of positive restructuring. It can be argued that real restruc-turing in terms of the locus of control over employment change was not the objective of such efforts. If one large company was in difficulty in a community, the natural tendency was to look for another to replace it. Salvation was to be sought from the outside.

The community development approach during this early period was frequently marginalized. It was often associated by many people with the type of community development that was focused on social development or, if there was an economic thread to it, with a collectivist approach to economic development (MacLeod, 1979). In the former, social develop-ment was certainly an important objective to pursue, but without incorporating or integrating with a significant economic development thrust, community (social) development easily falls into a syndrome of dependency, of dealing with the symptoms of economic malaise and only with marginalized groups in the community.

On the other hand, combining social and economic development objec-tives in a broader community-based approach to development seems to offer greater scope for significant restructuring of employment, including attempting to bring marginalized groups back into the mainstream. The opportunity is there through a variety of means, such as small business development, employee buy-out or takeover and collective approaches to community enterprise, to engage in positive restructuring of rural employ-ment opportunities. This should not be taken to mean that Canada's rural communities, especially those in the rural peripheries, can expect significant growth in employment. Stabilization, limiting decline and restructuring in a development process present a more realistic range of alternative objectives for many of the communities involved. Their efforts still have to be cast into some form of long-term framework and, above all else, be tempered with realistic analysis of goals, opportunities and constraints.

Relationships with Upper Levels of Government

The complex mosaic of approaches to community development in rural areas and small towns in Canada is complicated by relations with upper levels of government (Bryant, 1989), which have created enabling environments for community development that vary greatly from province to province (Bryant and Preston, 1987b; Savoie, 1986). However, there are still strong links to the past in terms of 'top-down' approaches.

There is a real need to foster capacity building at the local level rather than creating attractive 'straws' for communities to clutch at but which ultimately leave communities high and dry. For example, Atlantic Canada has experienced a long history of 'top-down' approaches; and even with more decentralized institutions, such as Enterprise Cape Breton (created in 1985), the focus remained on large-scale development projects and investment from the outside, projects that experienced a much higher rate of failure than the smaller-scale projects. In July 1991, it was replaced by Enterprise Cape Breton Corp. which has a mandate to focus on growth and development from within; only time will tell whether there is a real shift.

By definition, upper levels of government cannot undertake community development. Upper levels of government do have roles to play though in creating favourable enabling environments. Increasingly concerned for efficiency and effectiveness in the utilization of public funds and in safeguarding broader collective interests, a major area of activity for upper levels of government is in the area of partnership arrangements with communities. This is easier said than done, for true partnerships require sharing in management and risk-taking, and evaluation of such arrangements necessarily requires evaluating the upper levels of government as well as the local community role. Other roles for upper levels of government are in education, local capacity-building and information gathering, analysis and transfer.

There have been a number of relatively positive experiences in the community economic development field in Canada (see Bryant and Preston, 1987b; Pasadena, 1988), but perhaps the one that best exemplifies the potential for communities to take charge of their own affairs in a productive relationship with upper levels of government is the 'Community Futures programme' administered by Employment and Immigration Canada. Aimed at communities in relatively disadvantaged areas, it is based on intercommunity cooperation and long-term planning. In its short existence, it has probably had a positive impact on attitudes towards community development that surpasses any other federal or provincial programme in the country. The programme is not expensive, and is focused largely on the local human resource base in terms of organization, implementation and small business development.

Conclusions

During the 1980s in Canada, there have been significant changes in the nature of local community economic development efforts, principally moving towards approaches that can be labelled 'sustainable community development'. They involve greater integration of social and economic objectives, a relatively broadly based community input into the planning and implementation process, and a greater focus on change from within the community. For success, they require: a realistic long-term planning framework, linked to action and oriented towards the future and not simply to resolving current 'problems'; a relatively decentralized approach to responsibilities; and effective partnerships or team-building, including intermunicipal, intergovernmental and private/public sector cooperation.

The local development scene is, however, still dominated in most regions by strong ties to the industrial development paradigm. In some respects, this represents a misreading of the changing nature of systems of exchange. It is argued that the community development process will have to move further towards (sustainable) community development approaches. A major requirement will be to ensure the creation of more favourable enabling environments at provincial and federal levels as responsibilities for governance are increasingly decentralized. This is likely to happen due to: (i) considerations of effectiveness and efficiency in the use of public resources; and (ii) greater demands for involvement by local populations in shaping their local environment. A sound community development process appears, therefore, to be essential in managing rural employment restructuring so that rural areas can rebuild their vitality and reinforce their role in the national economic system. This process, while community-based, must be outward looking and this requires communities to undertake a realistic appraisal of where they and their actual and potential economic base fit into the various systems of exchange.

References

Bell, D. (1973) *The Coming of Post-Industrial Society: A Venture in Social Planning*. Basic Books, New York.

Bryant, C.R. (1980) Manufacturing in rural development. In: Walker, D.F. (ed.) *Planning and Industrial Development*. Wiley, London, pp. 99–128.

Bryant, C.R. (1988) Economic activities in the urban field. In: Russwurm, L.H., Coppack, P.M. and Bryant, C.R. (eds) *Essays on Canadian Urban Process and Form III: The Urban Field*. Publication Series 30, Department of Geography, University of Waterloo, Waterloo, Ontario, pp. 57–79.

Bryant, C.R. (1989) Entrepreneurs in the rural environment. *Journal of Rural Studies* 5, 337–48.

Bryant, C.R. (1991a) Rural community, land use dynamics and sustainable development. In: van Oort, G., van den Berg, L.M., Groenendijk, J.G. and Kempers, A.H.H.M. (eds) *Limits to Rural Land Use*. Centre for Agricultural Publishing and Documentation (Pudoc), Wageningen, The Netherlands, pp. 9–15.

Bryant, C.R. (1991b) Les grandes région métropolitaines au Canada: les choix du 21e siècle. *Etudes Canadiennes* 29, 7–18.

Bryant, C.R. (1991c) The service industries in Ontario. In: Mitchell, B. (ed.) *Ontario: People, Economy and Environment*. Publication Series 34, Department of Geography, University of Waterloo, Waterloo, Ontario, pp. 63–84.

Bryant, C.R. and Coppack, P.M. (1991) The city's countryside. In: Bunting, T. E. and Filion, P. (eds), *Canadian Cities in Transition*. Oxford University Press, Toronto, pp. 209–38.

Bryant, C.R. and Johnston, T.R.R. (1991) *Agriculture in the City's Countryside*. Pinter Publishers, London and University of Toronto Press, Toronto.

Bryant, C.R. and Preston, R.E. (1987a) A framework for local initiatives in economic development. *Economic Development Bulletin* 1, Economic Development Program, University of Waterloo, Waterloo, Ontario.

Bryant, C.R. and Preston, R.E. (eds) (1987b) Local initiatives in economic development. *Papers in Canadian Economic Development* 1, University of Waterloo, Waterloo, Ontario.

Bryant, C.R. and Preston, R.E. (1987c) Strategic economic planning and local development. *Economic Development Bulletin* 2, Economic Development Program, University of Waterloo, Waterloo, Ontario.

Bryant, C.R. and Preston, R.E. (1990) Economic development in small town and rural environments. *Economic Development Bulletin* 10, Economic Development Program, University of Waterloo, Waterloo, Ontario.

Bryant, C.R., Preston, R.E. and Dudycha, D.J. (1988) Marketing and local economic development. *Economic Development Bulletin* 5, Economic Development Program, University of Waterloo, Waterloo, Ontario.

Coffey, W.J. (1987) Structural changes in the Canadian space economy, 1971–1981. In: Coffey, W.J. and Polèse, M. (eds) *Still Living Together: Recent Trends and Future Directions in Canadian Regional Development*. The Institute for Research on Public Policy, Montréal, pp. 73–120.

Coffey, W.J. and Polèse, M. (1985) Local development: conceptual bases and policy implications. *Regional Studies* 19, 85–93.

Coppack, P.M. (1985) A stage model of central place dynamics in Toronto's urban field. *East Lakes Geographer* 20, 1–13.

Decter, M.B. (1989) Diversification and single industry communities: the implications of a community economic development approach. *Local Economic Development Paper* 10, Economic Council of Canada, Ottawa.

Douglas, D. (1989) Community economic development in rural Canada: A critical review. *Plan Canada* 29, 28–46.

Dykeman, F.W. (1989) Local rural planning and development in Atlantic Canada: Perspectives and direction. *Plan Canada* 29, 99–107.

Dykeman, F.W. (1990) *Entrepreneurial Communities*. Rural and Small Towns Research and Studies Programme, Mount Allison University, Sackville, New Brunswick.

Intergovernmental Committee on Urban and Regional Research (ICURR) (1990) *Community Development: An Interprovincial Forum.* Ontario Ministry of Municipal Affairs, Toronto.

Joseph, A.E., Keddie, P.D. and Smit, B.E. (1988) Unravelling the population turnaround in rural Canada. *The Canadian Geographer* 32, 17–30.

MacLeod, G.J. (1979) *Community Development Corporations: Theory and Practice.* Bulletin 7, New Dawn Enterprises Ltd, Sydney, Nova Scotia.

O'Neill, T. (1990) *From the Bottom Up: The Community Economic Development Approach.* Economic Council of Canada, Ottawa.

Pasadena (Economic Development Committee) (1988) *Awakening Entrepreneurial Spirit: The Key to Job Creation.* Pasadena Economic Development Committee, Pasadena, Newfoundland.

Perry, S.E. (1989) *Regional Development through Community Involvement.* Economic Council of Canada, Ottawa.

Savoie, D. (1986) *Regional Economic Development – Canada's Search for Solutions.* University of Toronto Press, Toronto.

Stohr, W.E. and Taylor, D.R. (1981) *Development from Above or Below.* Wiley, New York.

19

HOME-BASED BUSINESS: OPPORTUNITIES FOR RURAL ECONOMIC DIVERSIFICATION AND REVITALIZATION

Floyd Dykeman

Rural Community Stresses and Their Management

Rural communities in Canada, as in most developed countries, face very serious economic problems. Resource industries, often the economic mainstay of rural communities, are undergoing significant restructuring relating to global and national economic interdependence and linkages, and technological innovations. Farmers are increasingly having to seek off-farm employment or diversify farm activity; the family farm is being replaced by corporate farming that is large-scale, capital-intensive and technology-oriented, thus reducing employment opportunities; competition from Third World forestry operations threatens world markets for forestry operators in Canada, as do considerations of resource quality and quantity. Resource supply problems in the fisheries sector, often attributed to foreign fishing fleets and their use of modern technology, contributes to fish plant closings and massive labour force disruption in small communities; and the closing of mine operations because of ore quality, the costs of extraction, or mineral depletion means massive lay-offs. These changes have all contributed to economic adjustment and restructuring for rural communities in Canada. Technology, changing demographics including an ageing population, more single parents and smaller families, government policies such as free trade, GATT talks and changing societal attitudes toward the environment generally add to the complexity of defining, understanding and managing rural communities.

The problems in addressing the forces of change that affect rural Canada are generally compounded by rural regions that have traditionally had lower per capita incomes, persistently high unemployment rates and dependence upon resource industries such as forestry, mining, fishing and agriculture, with limited processing of the raw materials extracted from the region. Furthermore, their lack of competitiveness has been affected by factors such as lack of entrepreneurship, poor management, slow adaptation to technology, high illiteracy rates and poor labour force skills.

Stress Management for Rural Communities

The response to regional underdevelopment issues in Canada provides a rich history of almost 30 years of concentrated efforts by federal and provincial governments (Savoie, 1986). Although the lessons that can be gleaned from this experience are many and valuable, increasingly local leadership initiative through rural and community development is emerging as a potentially productive means of managing rural communities (see Chapter 18).

Through a variety of community-based economic initiatives, rural communities in Canada are developing or reorienting local institutions. The aim is to identify and implement strategies to assist communities with realizing goals related to social and economic diversification and development, as well as environmental management. This community-based economic development approach is not particularly new. It involves local leaders taking an initiative to identify local endogenous resources, and assessing how those resources can be used most effectively to regenerate the community and contribute to residents' quality of life. This approach is often achieved by levering resources through strategic partnerships forged among communities within a region, between communities and the private sector, and between communities and upper levels of government.

While traditional economic development initiatives focused on attracting industry to the community, recent community-based development approaches focus on building upon existing resources and strengths from within, through an integrated strategic planning and development approach. Inevitably, community-based development strategies vary but usually involve one or more of the general strategic development initiatives outlined in Fig. 19.1. Of particular significance to work being undertaken by local rural economic development leaders is a focus on entrepreneurship and small business development and on developing or creating an enabling environment that is supportive of entrepreneurs and small business (Bryant, 1989).

Self-employment and Small Business Development: Prevailing Development Forces

Canada's commitment to small business is justified, based on past performance regarding job creation. In Canada during the period 1975–1982, the greatest job creator in relative terms was small business which accounted

Fig. 19.1. Strategic development initiatives in community-based development.

for 76% of the growth, and all of this growth occurred within young small businesses that employed between one and four persons (Atlantic Consulting Economists, 1988). The Royal Commission on Economic Union and Development Prospects for Canada indicated that, between 1978 and 1992, 66% of the employment growth in Canada in all sectors was accounted for by small business growth. Small business is clearly a major stimulant to the national, provincial and community economies.

Canada's performance regarding self-employment is also impressive. As reported by Statistics Canada (1988), the number of self-employed increased more than twice as fast as the number of paid workers. The number of self-employed women increased three times faster than for men. Statistics Canada concluded that one out of every seven workers was self-employed in 1986. Their study also indicated that the self-employed make up a smaller proportion of non-agricultural employment in Canada's metropolitan centres (10.5%) than they do within the smaller urban centres and rural areas (13.4%). Self-employment, therefore, is important to the local economies of small town and rural Canada.

More recent work by the Atlantic Canada Opportunities Agency (ACOA) (1991) noted that one-third of the employment in Atlantic Canada was employed by small business, and small business within the region accounted for 75% of new jobs created from 1979–1988. While small business growth is impressive, large corporations still account for the majority of jobs. ACOA also notes that micro-businesses, those employing less than five employees, are more prevalent, accounting for 78% of the businesses employing fewer than 100 employees. These micro-enterprises also accounted for 49% of the new jobs created in the Atlantic region over the 1979–1988 period.

Therefore, it can be safely concluded that small business is a prevailing economic force nationally and regionally and is important in a rural context. The federal government's emerging interest in small business development, the release of a federal government policy on entrepreneurship, and the formulation of new policies and programmes by the provincial governments across Canada all attest to the public policy attention focused on this component of the economy.

Entrepreneurship, Small Business Formation and Rural Development

The numbers, although crude, suggest that entrepreneurship and small business formation need to be considered seriously as components of a rural community development strategy. However, the nature of small business development and rural communities suggests that a cautious and well-researched approach to this type of strategy is needed. In other words, entrepreneurship and small business development, although an

Table 19.1. Constraints faced by small business in rural communities.
***
- Small population base
- Often a declining or stable market area
- Lower education levels
- Inadequate workforce skills
- Limited access to economic and technical information meaning fewer innovations
- Limited capital available because of higher risks
- Less wealth within communities and lower disposable income
- Limited opportunities for business networking because of population and business isolation
***

alluring strategy for economic diversification, may encounter difficulties. For example, Table 19.1 lists typical constraints that rural areas, by their very nature, pose for entrepreneurs and small business. These constraints can affect competitive positioning significantly.

Many communities have small business and entrepreneurial development opportunities that can be exploited. Opportunities become available as new ways of producing goods are developed, new markets are opened, or new ways of using existing resources are discovered. For most communities there are reasons for optimism.

Research by the National Governors' Association (1988) in the US found that owners of new rural businesses tend to be 'home grown'. Often the business is started in the owner's home town, or within commuting distance of the home town, and approximately two-thirds of entrepreneurs start businesses in communities where they were born or raised.

Using entrepreneurship and small business development as part of a rural development strategy involves risks. For example, small business failures generally are high and, given the nature of rural areas, rural enterprises are often small and therefore fragile (OECD, 1986). However, on a more positive note, Popovich and Buss (1989) found that new businesses started in North Dakota since 1980 employed nearly a quarter of the State's workforce, both in urban and rural areas. Two-thirds of these businesses survived through to 1987; most businesses were home grown; most started small and grew; and most small businesses provided acceptable incomes, although the majority of them supplemented the family income.

As rural communities seek to diversify and revitalize their economies, the arguments for supporting and encouraging entrepreneurship and small business are sufficiently compelling to take the opportunities presented seriously (Table 19.2). Research also shows that towns with a strong small business and entrepreneurship base also have good retail facilities, a higher level of home ownership, fewer slums, better sanitation standards, and higher expenditures on education, recreation and religious activities.

Table 19.2. Arguments for supporting and encouraging entrepreneurship and small business.

- Contributes to local growth/development
- Creates jobs
- Inspires innovation among existing businesses
- Positive impact on community stability through diversification
- Contributes to higher incomes within communities
- Results in a higher level of civic participation by industry leaders
- Locally controlled enterprises with a concern for community

---

Source: adapted from Kent (1984).

Branch plant and external industry location within rural communities should not be ignored as a development strategy. However, entrepreneurship and small business development can contribute to locally controlled enterprises. This is important because local business people are more likely to make decisions that are sensitive to the community. Such decision making adds a further element of stability and contributes to a sustainable future for the community. As Shapero (1984) notes, communities' successes with encouraging small business formation beget small business developments.

Understanding Home-based Business as a Rural Diversification Opportunity

Coffey and Polèse (1984) have suggested a stages model of endogenous regional and community development that starts with the emergence of local entrepreneurship and local firms. Small business strategies pursued by rural communities fit this model. There is much more research to be undertaken to develop an adequate understanding of rural entrepreneurship and small business development, and one of the critical elements concerns a very fundamental form of small business – home-based business.

Trends favouring home-based business development

The location of business activity within the home is not a new phenomenon. Prior to the industrial revolution, much economic activity was home-based. The advent of the industrial economy meant that the work was mechanized, requiring a factory setting for work activity. This trend toward industrialization substantially reduced the importance of the home as a workplace (McRae, 1988). After a century of emphasis on separating business from the dwelling, societal changes are evident that are conducive to the emergence of the home-based business once again (Table 19.3).

Table 19.3. Societal change supporting the emergence of home-based business.

- A shift in values regarding work and quality of life reflecting an increasing desire for independence and flexibility
- Movement from an industrial to an information economy
- Availability of technology enhancing long-distance communication data, reports, etc., from the home
- Decentralization of production systems
- Increasing desire for customized products and services
- Changing attitude of the consumer demanding uniqueness, variety, and quality
- Changing family structure with an increasing number of single-parent families

Sources: Plummer (1989), Hawken (1983), Naisbitt (1982) and McRae (1988).

These trends make working from the home both attractive, convenient, and potentially profitable. However, the economic advantages of operating a small business from the home cannot be ignored. The home offers a rent-free location for new business start-up, thereby reducing overhead costs and providing the entrepreneur with a competitive edge. Use of the home as the location for business also helps reduce the initial investment required to start the business and, therefore, reduces the upfront risks. The home in this instance serves as an incubator for small business which can be critically important in rural settings where space for business is often at a premium.

Christensen (1988) identifies five factors that contribute to self-employment in the home. These are: (i) the desire to go into business for oneself; (ii) a flexible work environment, particularly for women with child-care responsibilities; (iii) retirees wanting to supplement the pension; (iv) older women seeking to re-enter the work force; and (v) individuals who are the victims of lay-offs seeking new employment options.

Problems faced by home-based business

Any strategy that uses home-based business as part of a rural community economic diversification programme needs to recognize that these businesses face many of the start-up and maintenance problems associated

Table 19.4. Problems faced by home-based business.

Community-based
- Land use changes/impact
- Nuisance to neighbours
- Additional strain on municipal services
- Loss of commercial tax revenue
- Formal business community sees unfair competition

Family-based
- Conflicts between home and business spaces
- Conflicts between family and business patterns

Operating
- Access to financing and venture capital
- Lack of time to market and promote
- Need for networking
- Lack of management skills

Public policy issues
- Transfer of corporate costs
- Taxation policy
- Labour, health, and safety laws
- Child care laws

with small business generally. In addition, they face other problems, many of which are specific to this micro-enterprise.

These problems fall into four categories (Table 19.4) and reflect problems with community, problems associated with family and the sharing of home space for both work and family, operating problems, many of which are common to small business generally, and public policy issues that arise directly as a result of this business type. All of these need to be carefully understood and dealt with if home-based businesses are to be used as a tool for rural community economic diversification.

Incidence of Home-based Business

Overview

Information on the numbers of home-based businesses varies considerably. As one popular US magazine announced: 'there is no doubt about it: home-based business is booming!' (*Entrepreneur*, 1990, p. 76). The magazine estimated that there were 18.3 million home-based businesses in the US in 1989; no breakdown for urban and rural distribution exists, but this was a 23% increase over 1988. A survey in the US in 1984 indicated that 13% of the households had some business working from the dwelling, and that 26% of the workforce worked from the home on a regular basis (National Alliance of Home-Based Businesswomen, 1987).

A recent study in Prince Edward Island on the cultural industry indicated that 60% of the industry operates from a home base (The DPA Group, 1988). Other sources have suggested that 50% of all small businesses start in the home and that as much as 70% of businesses led by women start from the home (Government of British Columbia, 1988). The British Columbia government has indicated that there are 12 000–15 000 individuals per year who launch business ventures from their home in the province. Although the data are not complete, they suggest that home-based businesses are common.

A survey of home-based businesses by the Okanogan County Council for Economic Development, Washington State, noted that in a rural county of 31 000 population there were 540 home-based businesses collectively responsible for US$6.5 million in gross sales in 1985. These home-based businesses provided 149 full-time jobs, 257 part-time jobs, and 118 seasonal jobs. A total of 48% had operated for more than 5 years, 25% derived a portion of their income from the business, and 10% derived all their income from the business (Johnson-Rodriguez, 1987).

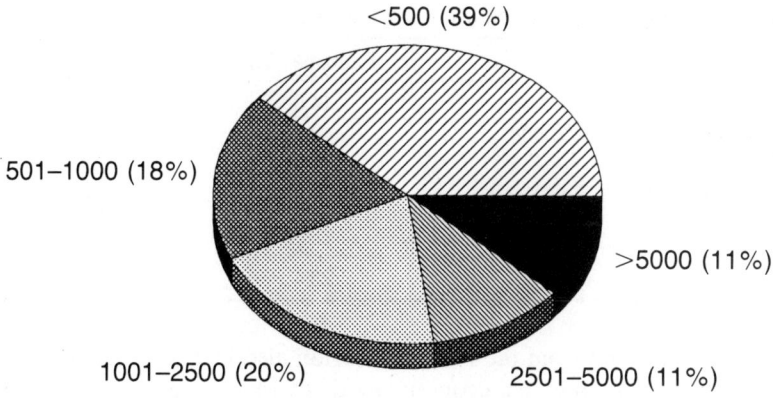

<500 (39%)

501–1000 (18%)

>5000 (11%)

1001–2500 (20%)

2501–5000 (11%)

TOTAL BUSINESSES

Fig. 19.2. Distribution of home-based businesses by community size (population) in Atlantic Canada.

The Atlantic Canada situation

Based on a survey of rural and small town residents of Atlantic Canada (Strople, 1987), it was estimated that 9.6% of the homes in Atlantic Canada maintained a home business. About 42% of these were part-time businesses (involving less than 20 h per week), while the remainder were full-time businesses. In rural areas, 12.3% of the homes maintained a business, and in urban areas 7.2% of the homes maintained a business.

Using Statistics Canada's data on the number of homes in the region and the survey results, it is estimated there were 63 000 home-based businesses within the region. Assuming that each home-based business operating full-time created one job, these involved about 37 000 full-time jobs.

A very high concentration of these home-based businesses are located in small communities, especially in communities in the 'less than 500' population-size category (Fig. 19.2). A majority of the businesses are found in the retail and services sector (Table 19.5). (In this case the 'other' category represents retail services, such as hairdressers.) In rural areas, there are slightly more businesses involved with the resource sectors of agriculture and forestry. An issue not well addressed by the data is the distinction between farming, on the one hand, and home-based but farm-related businesses that add value to the primary agricultural product on the other. This issue must be addressed in subsequent, more focused research on home-based business.

Table 19.5. Types of business: part-time and full-time.

	Part-time (%)	Full-time (%)
Agriculture/forestry	14.9	18.0
Retail (equipment and machinery)	3.0	10.0
Retail (other)	26.9	24.0
Light manufacture	10.4	2.0
Professional services	9.0	5.0
Technical services	13.4	14.0
Other	22.4	27.0

The survey data from the Atlantic Region also highlight the significant contribution to the regional economy made by revenue generated through home-based business. Based on the incidence of home-based business in the survey, the number of homes within the region, and the mid-point of the ranges for gross income used in the survey instrument, it is estimated that the home-based business component of the region generated C$1.2 billion revenue in 1985. The survey also shows that the majority of the businesses had gross incomes of less than C$10 000 (53%), whereas almost 20% had gross incomes exceeding C$50 000 (Fig. 19.3). Rural areas were more likely to have incomes greater than C$50 000 than urban-based businesses. Among part-time businesses, 55% had incomes that were less than C$2500, compared with 8.9% of full-time businesses.

Therefore, while home-based businesses are small and do not have a

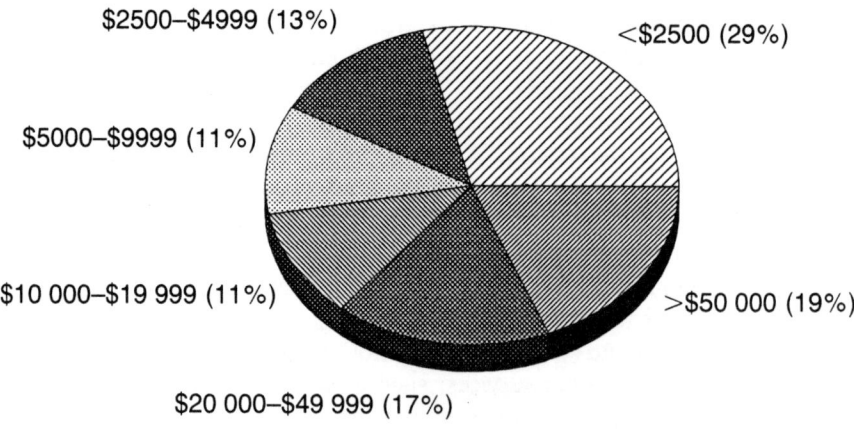

$2500–$4999 (13%)
<$2500 (29%)
$5000–$9999 (11%)
$10 000–$19 999 (11%)
>$50 000 (19%)
$20 000–$49 999 (17%)

REGIONAL

Fig. 19.3. Distribution of home businesses by income category (C$) in Atlantic Canada.

high public profile within the region, their importance to rural and small community economies, as well as to the regional economy, cannot be overlooked when developing employment and small business development strategies and programmes.

Sackville: a small town example

Building on the data presented for 1985–1986, further work was undertaken in 1990 by the Commerce Department, Mount Allison University, focusing on the Town of Sackville (population 5000). Based on three methods of identifying home-based business (identification is the most difficult methodological issue with home-based business research), a total of 89 home-based businesses were identified, accounting for about 5% of the homes in the study area. This number, however, is regarded as low by the researchers, and ongoing work is being undertaken to assess better methods, or combinations of methods, to identify reliable home-based business lists.

Most of the home-based businesses in Sackville had been established for some time, with 12% of the businesses having operated from the home for 15 or fewer years, slightly over 50% for less than 10 years, and 28% for 5 or fewer years. Few home-based businesses employed other members of the family; 50% had no other members of the family employed, whereas 39.5% had one or two members employed.

In 32% of the home-based businesses all of the business transactions occurred locally; 19.1% of businesses had 10% or less of their sales locally, suggesting that these constituted the 'export-oriented' (that is, community export) businesses that bring new dollars into the community, thereby generating new wealth. From an economic development and diversification perspective, these businesses are important.

Of the 'export-oriented' businesses, 45% had less than C$15 000 gross revenues in 1989; 12.9% had gross revenues of C$15 000–30 000; 14.5% had revenues of C$30 000–50 000; and 14.5% had revenues of C$50 000–100 000. Most respondents had capacity for increasing business, and 25.9% of the home-based businesses desired a 25% or more increase in business activity.

For a small community like Sackville, a number of conclusions need to be given serious consideration as the community seeks to plan for a diversified economy. These are: (i) home-based business constitutes an important opportunity when preparing a business diversification strategy for Sackville; (ii) the 'export' businesses are those that will help to produce new wealth for the community and should be carefully nurtured to continue job growth and development; and (iii) the community should develop an understanding of home-based business support needs among those businesses interested in and planning expansions.

Rural Economic Diversification Using Home-based Business as a Catalyst

It is clear that small business is a growing component of the Atlantic Canada regional economy. It is receiving considerable attention as a catalyst for regional development, growth and progress. There is also a growing movement by rural communities within the region to demonstrate leadership and entrepreneurial spirit; these qualities are required to assess strengths and weaknesses realistically as a basis for identifying opportunities that can contribute to enhanced rural community welfare. Clearly the challenge lies in addressing the three building blocks of 'community' – economic, social and environmental.

Most small communities in Atlantic Canada depend economically on one or two large employers; this places the community in a potentially economically vulnerable position. Many employers are resource industries involved with the agriculture, forestry, fishery and/or mining sectors. These industries are themselves undergoing significant restructuring, resulting in large lay-offs of local employees and, in some cases, plant closures. As these small communities undertake a self-examination, they are quick to realize that economic diversification is essential. For rural communities, economic diversification means development and growth of business and industry that is usually small in scale and, for the most part, locally owned. Furthermore, this type of economic activity functions in non-traditional sectors of the community, or in the traditional sectors but by adding a new dimension to the production activity, therefore contributing to community sustainability.

Home-based business offers a significant opportunity to use local entrepreneurial spirit; this enables a business base to be established within the community, thereby contributing to community well-being and diversification through the provision of new jobs and increased incomes. However, proactive initiatives on the part of the community must be addressed if success with diversification is to work, especially using home-based business as a catalyst. Three strategies need to be addressed: (i) the development of a retention and expansion strategy; (ii) the development of a home-based business start-up support strategy; and (iii) the development of a land-use and development control policy at the municipal level, supportive of home-based business.

Retention and expansion

Essential for any initiative of this kind is the formation of an intelligence and information system that allows a rural development officer the opportunity to track progress among existing home-based businesses within the community. Once this system is in place, an on-going programme of visits, business planning and management support is essential to assist the home-based business owner. Such support should be aimed at maintaining the economic health of the small business, as well as developing the possibility of expansion. The information system provides a development officer with facts giving insight into the problems typically faced by this business type and, therefore, insights into the appropriate response of a local community development officer.

Start-up support

In many communities there are individuals who are unemployed and contemplating self-employment. Many of these individuals will use the home as the initial site for a business start-up. It is important for a local development officer to provide needed assistance with start-up to enhance the opportunities for success. Help will involve assistance with business planning, financing, management, marketing and, possibly, sales. Start-up support may also involve providing the new, potential, home-based business entrepreneur with access to a network of other experienced home-based business entrepreneurs: they can provide encouragement, advice and other forms of needed support.

Land use policy

In many small communities, land use plans and by-laws have been adopted based on model by-laws used in other communities, or offered to the community by the provincial department responsible for community planning. These planning devices are often not sensitive to the need for allowing the location and operation of home-based business within small-community residential areas. Many communities have in fact adopted policies that counteract and even contradict one another. On the one hand, they encourage and actively pursue economic development and, on the other hand, they discourage home-based business through community plans and associated by-laws. Communities need to give greater thought to these local policy instruments and their impact on economic diversification options.

Conclusion

It is not difficult to compile a list of businesses that work from the home. For example, rural residents have chosen to open their farms to farm vacations, to operate bed and breakfast establishments, to make and sell Victorian women's clothing or other crafts on site, to use an abandoned shed as a machinery repair facility, to use an unneeded household room for a yarn shop, to establish a rural-based law or accounting practice in the home, or to contract services as a data entry clerk or word processor. Major international corporations have had modest starts as home-based businesses. Apple Computer, Ashton-Tate Software, Ford Motor Company, Gillette, Hallmark Cards, Hershey Foods, Nike and Readers' Digest serve as concrete examples of successful businesses that had home-based beginnings (*The Home Business Advocate*, 1988). Home-based business cannot be ignored as a legitimate rural development strategy. It is not, however, to be regarded as a strategy that should be pursued by rural communities in isolation of other strategic initiatives.

The opportunity to use home-based business as a rural economic diversification tool is considerable. It is an opportunity that requires, however, more focused attention by the rural development research community and rural communities themselves. Efforts that supply better information and understanding of home-based businesses are now needed, providing informed guidance for rural community economic diversification initiatives.

References

Atlantic Canada Opportunities Agency (ACOA) (1991) *The State of Small Business and Entrepreneurship in Atlantic Canada*. ACOA, Moncton, New Brunswick.

Atlantic Consulting Economists (1988) *Small Business and Job Creation*. Background report, Royal Commission on Employment and Unemployment, Newfoundland and Labrador.

Bryant, C.R. (1989) Entrepreneurs in the rural environment. *Journal of Rural Studies* 5, 337–48.

Christensen, K.E. (1988) Introduction: White-collar home-based work – the changing US economy and family. In: Christensen, K.E. (ed.) *The New Era of Home-Based Work: Directions and Policies*. Westview Press, Boulder, Colorado, pp. 1–29.

Coffey, W.J. and Polèse, M. (1984) The concept of local development: a stages model of endogenous regional growth. *Canadian Journal of Regional Science* 55, 1–12.

Government of British Columbia (1988) News Release. In: Ministry of Economic Development, *Provincial Government to Encourage Home-Based Business*. Victoria, British Columbia.

Hawken, P. (1983) *The Next Economy*. Ballantine Books, New York.

Johnson-Rodriguez, R. (1987) Economic development begins at home. *Home Business Network Newsletter* 1, Spokane, Washington.

Kent, C.A. (1984) The rediscovery of the entrepreneur. In: Kent, C.A. (ed.) *The Environment for Entrepreneurship*. Lexington Books, Lexington, pp. 1–19.

McRae, J.J. (1988) *Home-Based Business: Their Evolution and Potential*. A paper presented to the conference *Control Your Own Destiny: Home-Based Business in British Columbia.* Ministry of Economic Development, Prince George, British Columbia.

Naisbitt, J. (1982) *Megatrends: Ten New Directions Transforming Our Lives.* Warner Books, New York.

National Alliance of Home-Based Businesswomen (1987) *Planning for Homebased Businesses.* Midland Park, New Jersey.

National Governors' Association (1988) *A Brighter Future for Rural America.* Center for Policy Research, Washington DC.

Organisation for Economic Co-operation and Development (OECD) (1986) *Rural Public Management.* OECD, Paris.

Plummer, J.T. (1989) Changing values. *The Futurist* January-February, pp. 8–13.

Popovich, M.G. and Buss, T.F. (1989) Entrepreneurs find niche even in rural economies. *Rural Development Perspectives* 5, pp. 11–14. (United States Department of Agriculture, Washington DC.)

Savoie, D.J. (1986) *Regional Economic Development – Canada's Search for Solutions.* University of Toronto Press, Toronto.

Shapero, A. (1984) The entrepreneurial event. In: Kent, C.A. (ed.) *The Environment for Entrepreneurship.* Lexington Books, Lexington, pp. 1–19.

Statistics Canada (1988) *Enterprising Canadians: The Self-Employed in Canada.* Minister of Supply and Services, Ottawa.

Strople, G. (1987) *Perspectives on Community and Housing: An Overview of a Recent Survey of Rural and Small Town Residents of Atlantic Canada.* Department of Geography, Rural and Small Town Research and Studies Programme, Mount Allison University, Sackville, New Brunswick.

The DPA Group (1988) *Report on Economic Impacts and Opportunities in the Cultural Industry on Prince Edward Island. Executive Summary.* Communications Canada and the Prince Edward Island Department of Community and Cultural Affairs, Charlottetown, Prince Edward Island.

The Home Business Advocate (1988) Newsletter No. 12. Wendy Priesditz and Associates, Unionville, Ontario.

20

CHANGING THE PATTERN OF ABORIGINAL SELF-GOVERNMENT IN CANADA

Jackie Wolfe

Aboriginal Self-government and Land

In southern and central Canada, treaties established small reserves as places for Indian residence and formalized the process of non-aboriginal settlement on formerly Indian lands. Passage of the *Indian Act* promoted imposition of Indian Act band councils, replacing existing forms of governance. In the north and west historic treaties were not signed: nevertheless, Aboriginal peoples were dispossessed of their lands, and their governments were ignored or eroded when groups were coerced into taking up permanent residence in centralized settlements. In these and other ways, Aboriginal people have been detached from their lands and resource base, and from their governments.

Aboriginal self-government has, in the 1980s, become the prime agenda item for national, regional and local Aboriginal organizations (Erasmus, 1989). The agenda involves re-establishment of Aboriginal government over Aboriginal people, land and resources (Little Bear *et al.*, 1984; Canadian Arctic Resources Committee, 1988; Cassidy and Bish, 1989; Cassidy, 1991). Consequently, they are asserting facets of government which they regard as important, over their members and over the lands for which they have recognized jurisdiction. They are also working to exercise renewed authority for lands and resources, and policies and practices affecting their use, over which they do not presently have full or formally recognized jurisdiction.

> Indian people assert that their rights flow from their relationship with the land. Land is thus a prerequisite for and vital to self-government
>
> (Penner, 1983, p. 105)

When Aboriginal people use the term self-government they mean that they must define the powers they will exercise, and they will design decision-making structures appropriate to their situation and congruent with their needs and their culture (Wolfe, 1989). It is not regarded as an end in itself: rather it is a necessary step to bring about constructive and positive change for Aboriginal people in Canada. According to the former National Chief of the Assembly of First Nations: 'We are not trying to get recognition to govern ourselves, to be self-determining, as an "end" goal . . . We have all come to the conclusion that we can do the job better ourselves' (Erasmus, 1988, p. 51).

Functional self-government does not exist in a vacuum. It has a number of requisites. Four are particularly important: legitimacy, recognition, decision-making authority, and adequate and appropriate resources. Legitimacy, in this context, means that the government is supported by and has the respect of its constituents, and is recognized by them as being properly and legitimately constituted. The First Nations' leaders are working with their peoples to develop new forms of governance that have local support. Some are re-asserting traditional structures and processes for social control and resource allocation. Others are adapting the imposed forms to their own needs. To function effectively a government must also be recognized as an authentic government by those external to it, with which it must interact. Consequently, many of the First Nations are engaged in protracted negotiations with the federal and provincial or territorial governments. A government must also have decision-making authority for matters that are of central importance to its own people. First Nations are articulating their own priorities and are negotiating with, manipulating, or circumventing external governments in their bid to make their own decisions. Finally, to be effectively self-governing, a government must have sufficient resources to implement its decisions. This means having financial resources, including some measure of budgetary control, and having the human resources and competence to carry through with decisions (Wolfe, 1991).

Central governments have, until recently, had a limited perspective on Aboriginal government, favouring modifications of a municipal model of elected councils with devolved and limited authority to the exclusion of other more substantive changes (DIAND (Department of Indian Affairs and Northern Development), 1982; INAC (Indian and Northern Affairs Canada), 1984). Such devolution is variously termed self-government or self-management by central governments, but is not regarded as self-government by Aboriginal groups.

Aboriginal people in Canada live in four contrasting sociopolitical and constitutional contexts: the northern territories, which were not covered by historic land allocation treaties; the largely provincial north and mid-north beyond the settled zone but subject to land allocation through treaties, where the provinces own land and resources; the southern reserves in the close settled zones; and the urban places of the north and south. Patterns in greater self-government by Aboriginal communities are emerging in all but the latter (Fig. 20.1).

Paths to Greater Aboriginal Self-government

One categorization of pathways to greater Aboriginal self-government distinguishes between the greater self-government that many Aboriginal groups are exercising, whether or not formal legislation or sanctioned programmes exist (Cassidy and Bish, 1989), and those that are designed and promoted by central governments. Recognition of the former acknowledges that power shifts as much when a First Nation takes over control of child welfare, or education, or policing, gains a co-equal voice in land and resource co-management, or successfully negotiates to exercise authority over priorities it sets, as when they are delegated more functions of government. Emphasis on the latter reinforces the perception that the only changes that count are formalized changes made by external governments.

First Nations Insistence on New Approaches

Substantial shifts are taking place in the authority, as well as structures and functions of Aboriginal governments. Many of these are driven by Aboriginal initiatives at the community level. All have common threads: reassertion of Aboriginal government over Aboriginal people and land under Aboriginal jurisdiction; and exercise of an Aboriginal political voice with respect to land and resource use and environmental quality in areas where they do not have full and formal jurisdiction.

Constitutional affirmation of the right of self-government

Many Aboriginal leaders continue to argue that constitutional change incorporating the inherent Aboriginal right to self-government is the only guarantee of change to the established order. They see it as the only protection from unilateral repeal or amendment by later federal governments. Constitutional amendments failed with the fourth and final constitutional conference on First Nations self-government in 1987, and no new conference has yet been called. Constitutional affirmation continues to be an Aboriginal priority. Aboriginal leaders are now insisting that such constitutional affirmation be independent of, and not within, the context of the Canadian Charter of Rights as required by the federal government's 1991 constitutional reform proposals.

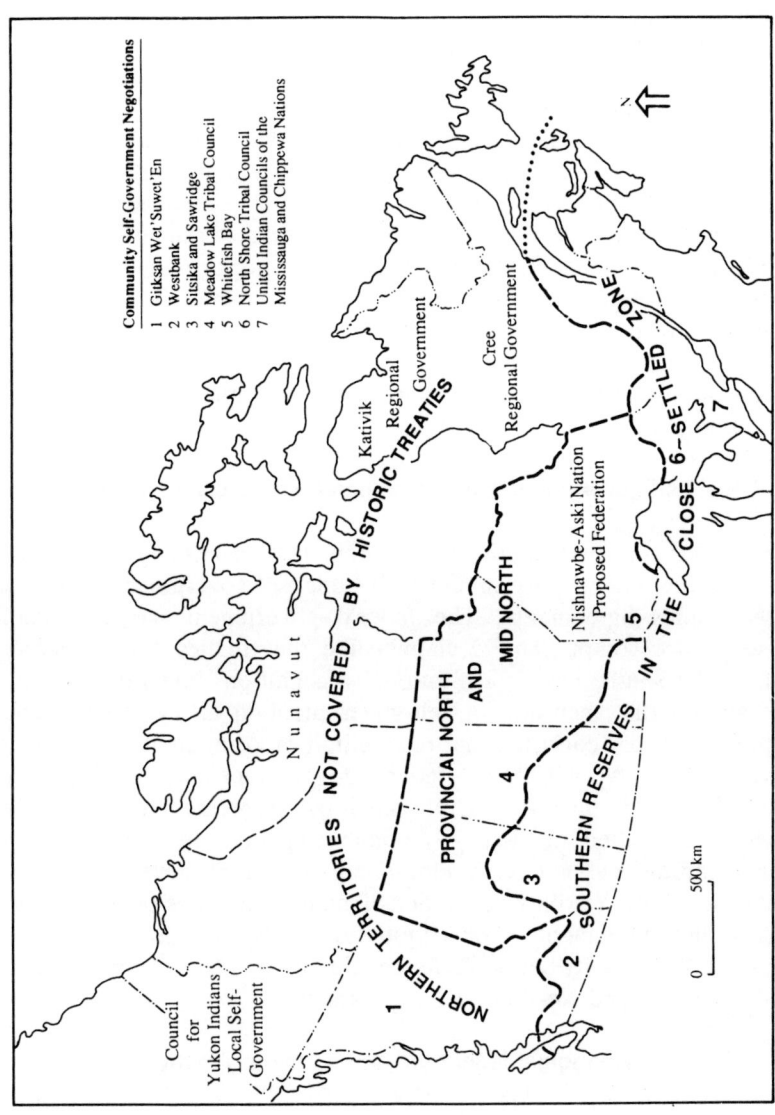

Fig. 20.1. Patterns of Aboriginal self-government in Canada.

Aboriginal public government

The 'most comprehensive new approach to indigenous self-government is the projected Nunavut Territory' (Jull, 1991, p. 16). The Inuit Tapirisat of Canada, representing the Inuit of the eastern and northern Northwest Territories (NWT), proposed division of the NWT and establishment of Nunavut in the late 1970s. The proposal was designed to create a jurisdiction in which Inuit culture and tradition could thrive by virtue of an overwhelming Inuit majority population (Nunavut Constitutional Forum, 1985; Weller, 1990). Establishment of Nunavut would put in place a public government in which the 13 000 or so Inuit of the eastern NWT would form over 90% of the population. The public government of Nunavut would, *de facto*, function as an Aboriginal regional government in the eastern arctic, an area approximately one-sixth of the Canadian land mass. It would correspond to the Nunavut comprehensive land claim settlement area. The federal government, territorial government and Tungavik Federation of Nunavut continue to support the principle of Nunavut. However, action has not yet been taken.

Local self-government within a comprehensive claims settlement

Aboriginal groups participating in comprehensive claims negotiations have pressured the federal government to include self-government as a clause in the claims agreements. The federal government has consistently refused, since comprehensive claims, like the treaties which preceded them in the south, have constitutional recognition. Federal negotiators have argued that inclusion of self-government in an agreement would set precedents for constitutional recognition of Aboriginal governments. However, on 1 April 1990 the federal government and the Council for Yukon Indians initialled an Umbrella Final Agreement (Crowe, 1990) which circumvented the issue by committing both the government of Canada and the Yukon government to negotiating self-government agreements with those Yukon First Nations (small local bands of Indians) that request such agreements. The negotiations will, though, be conducted within the Community Self-government Negotiations programme administered by Indian and Northern Affairs Canada.

Aboriginal federation and co-management

The Nishnawbe-Aski Nation (NAN) is an emerging federation of 46 Ojibway and Cree Bands or First Nations in remote northwestern Ontario, many though not all of whom have title to small reserves. The Nation's territory, most of which is presently held in provincial crownland, encompasses roughly half the province of Ontario. In much of the area

Nishnawbe-Aski peoples are in the majority and are the major direct users of the land. Where they are able, Nishnawbe-Aski bands continue to draw on traditional land-based resources for a substantial portion of their food and livelihood. Consequently, many of the chiefs feel that it is important to consolidate NAN control over the land before impacts from development further erode NAN use and *de facto* control (Nishnawbe-Aski Nation, 1991).

The governments of Canada and Ontario agreed to negotiate with the Nishnawbe-Aski Nation in February 1986 on 'the present and future self-government needs of NAN' (Nishnawbe-Aski Nation, 1986, p. 2). Since that time, the Federation has been developing a comprehensive framework for self-government. The framework makes provision for several types of governing institutions including: (i) aboriginal institutions for government of NAN lands and peoples and service delivery to them; and (ii) joint institutions, especially for land and resource planning and management and environmental protection.

NAN is proposing establishment of a Federation of Nishnawbe-Aski Nations. Within the Federation, each First Nation would be recognized as the primary level of Nishnawbe-Aski self-government and would govern its own internal affairs on its designated lands. The member First Nations will continue to be the most important level of Nishnawbe-Aski government, and all powers not specifically given to the Federation will rest with each First Nation (Nishnawbe-Aski Nation, 1991). The Federation will function as the regional authority for the Nishnawbe-Aski peoples and for the small and scattered parcels of reserve lands over which they have exclusive jurisdiction.

The NAN negotiating team has identified priority matters towards which they are targeting their self-government negotiations. These include policing and justice, child welfare and social services, land and resource planning and management, and environmental protection. NAN is working with the Ontario Provincial Police to establish a NAN Regional Police Service that is appropriate to the culture and traditions of the Nishnawbe-Aski people (Nishnawbe-Aski Nation, 1991). NAN already has three child welfare agencies, two of which function under provincial legislation. NAN proposes to expand its administration of social services to include general welfare assistance, family benefits allowance, homemaker and nurses services, family and individual counselling, and care of the elderly.

Negotiations on lands, natural resources and environmental protection pose particular challenges because the province of Ontario has jurisdiction over virtually all the lands and resources in question. NAN is proposing a threefold system of land classification and administration somewhat similar to that used in the James Bay and Northern Quebec Agreement (JBNQA) and Comprehensive Claims Agreements: namely, identification of reserve or community lands (Zone A) fully controlled by the First

Nations; First Nations lands (Zone B) over which each Nation would exercise specified authority for their own land use; and co-management lands (Zone C, which comprises the rest of northern Ontario) where NAN is proposing to share management powers and authority with the province. Development of co-management systems for Zone C lands is likely to prove particularly controversial and difficult to achieve, given both provincial and third-party forestry and mining interests.

NAN has a complex agenda to negotiate internally and with the federal and provincial governments (Ontario Native Affairs Secretariat, 1991). It accepts that the process will not be easy. But as a NAN negotiator observed: '. . . even if we got constitutional self-government tomorrow all of these things would still have to be worked out' (Bill Nothing, personal communication, 1991).

Land use and resource co-management

While the Comprehensive Claims Agreements have not made provision for direct forms of Aboriginal self-government, all contain numerous subagreements providing the Aboriginal claimants with a major role in land use planning, wildlife co-management and joint environmental management of those territories in which they have an interest but do not have absolute title. Even where final agreement has not been reached, many of these subagreements have been put in place. Aboriginal people have a 50:50 representation with government on management boards and commissions. Consequently, across the Yukon and NWT, Indians, Metis and Inuit are now in a position to exert considerable influence on land and renewable resource management.

Commentary

In the far north and mid-north of Canada patterns are emerging. First, Aboriginal governments, either in the form of ethnic governments or, as in the NWT, public governments with an Aboriginal majority, have greater than municipal powers, though less than provincial authority, over central places and lands immediately surrounding them. Second, regional groupings of governments have been established or are being proposed. These may be confederations of Aboriginal governments, or regional public governments, or both. However, these governments either have no jurisdiction or very limited jurisdiction for the vast lands lying between the central places that make up their membership. Nunavut is the exception in that it would have full jurisdiction over a contiguous land area. Third, signing of JBNQA in 1975 and, more recently, the Comprehensive Claims Agreements, gives Aboriginal people exclusive hunting, fishing and trapping rights to parts of the claims area, for which they have established or are developing management systems. Fourth, various land, resource and environmental management agreements with significant Aboriginal representation are in place or are being proposed for those lands between.

Within the Established Order

The federal government and INAC have established a number of programmes which provide opportunities to First Nations for greater self-government. These form a hierarchy from special Acts of self-government which provide First Nations with powers well beyond those possible under the *Indian Act* to lesser options possible within the Act.

Special self-government Acts

Although umbrella self-government legislation is not in place, special legislation has been enacted by the federal government for specific groups. The *Cree-Naskapi (of Quebec) Act* (1984) enabled the eight Cree bands covered by the *James Bay and Northern Quebec Agreement* (1975) (JBNQA) and the Naskapi of Schefferville covered by the *Northeastern Quebec Agreement* (1978) to take on a greater level of local self-government than that allowed under the terms of the *Indian Act* (Moss, 1985). In addition, the Cree local level band governments are complemented by a form of regional government operating through the Cree Regional Authority, and by special purpose bodies such as the Cree Regional Board of Health Services and Social Services, and the Cree School Board (Cassidy and Bish, 1989). Also under the JBNQA, another regional body, the Kativik Regional Government was established as a non-ethnic (though Inuit dominated) public government for the communities in the northern third of the province of Quebec. It administers, and provides delegated provincial functions for, the 15 Inuit villages in northern Quebec.

Community self-government negotiations

In the mid-1980s, a Self-government Negotiations Directorate was established within INAC. The Directorate operates a programme by which bands, tribal councils and other groupings negotiate special legislation enabling their councils to assume responsibilities and exercise authority beyond the terms of the existing *Indian Act*. The Directorate acknowledges that:

> Self-government negotiations will not alter the division of powers between the federal and provincial governments but will, through

practical measures, attempt to accommodate Indian government
within the existing constitutional framework.

(INAC, 1989, p. 1)

The minimum set of matters that may be dealt with includes: institutions
of government, membership, land title and land management, political
and fiscal accountability, continuation under the *Indian Act* or negotiation
of special legislation, and an implementation plan. The full range of
powers and functions can also include administration of justice, taxation
for local purposes, community services, education, health, social development, renewable resources, business and trades, and protection of the
environment (INAC, 1990).

Considerable interest has been expressed by Indian groups in the
community self-government negotiations programme. Currently, eight or
nine groups are working on so-called framework negotiations. Likewise,
eight or nine are at the next phase of substantive negotiations. However,
community self-government negotiations are lengthy. To date, none have
been finalized under the existing programme.

Alternative funding arrangements

Another INAC programme, Alternative Funding Arrangements (AFA),
allows band and tribal councils greater programme delivery flexibility
within existing legislation. The programme has four main features: greater
accountability of chief and council to the membership; optional multi-year
funding; local flexibility to transfer funds between programmes; and local
authority to modify or re-design federally funded programmes. So-called
self-management or self-administration is thereby increased.

The multi-year funding option makes possible funding agreements for
up to 5 years, and should improve the ability of councils to plan long-
term programmes with some certainty of the availability of funds. This,
combined with the transferability of funds, enables councils to establish
their own priorities. Councils have greatly expanded authority to develop
policies that reflect the needs and values of the local community by rede-
signing programmes and adopting suitable standards acceptable to the
community.

However, the Minister's overall accountability to parliament for funds
and programmes remains intact. Councils in AFA do not receive any
more funding than they would outside the programme. Further, bands and
tribal councils have to apply for entry into the programme, and eligibility
for the AFA programme is determined by senior regional departmental
staff. Bands in the AFA programme can use only such powers as are
delegated to them (INAC, 1986). Alternative Funding Arrangements are
little other than municipal-style government in a new package.

Commentary

The federal government has created a complex hierarchy of 'opportunities' for increased local self-government. INAC argues that the range of programmes provides for the varied needs of Aboriginal groups. Some groups are enabled to make substantial gains; others are relegated to lesser options, or withdraw entirely, thus intensifying the differentiation that already exists between those groups that have the capability to exercise additional powers and those that do not. Also, each group negotiates on a case-by-case basis, imposing high costs not only on the bureaucracies, involved but also on the communities and their leadership. At the present or even accelerated pace, it would take several decades for even a few of the 590 or so bands in Canada to achieve the more substantive forms of greater self-government.

Patterns and Implications

Across Canada a complex mosaic of Aboriginal governments is emerging, with several discernible patterns within it. The governments are a sometimes highly functional, sometimes uneasy mix, of traditional and non-Aboriginal elements. There are small local-level governments each with different powers and levels of authority, and different relationships with federal, provincial and adjacent Aboriginal and non-Aboriginal local governments. They have jurisdiction either over reserve lands (in treaty areas) or lands in fee simple absolute title (in comprehensive claims areas). A small though increasing number are developing governments that exercise authority over a range of functions that no municipality, however large, administers, let alone has policy responsibility for. They can be found in the most remote locations in the north and northeast, in the most urbanized southwest, and are being negotiated by several groups in more southerly close-settled areas.

Second, Aboriginal regional governments are in place or are proposed. These are found primarily in the more remote parts of the provinces and in the Territories. They are either ethnic governments, or public governments in which Aboriginal people form the overwhelming majority. Generally, they conform closely to a municipal or extended municipal model of government. With the exception of the proposed Nunavut government in the eastern arctic, they have jurisdiction only over the central places in the region, not the lands between.

Third, many First Nations are selectively taking on delivery of services to their people. When they take the initiative, and establish their priorities for what they will run, there is a clear ordering in their selection. The First Nations opt to control their own education, child welfare and other social services, health, policing and justice, and land and resource planning and environmental protection. They select those elements that they regard as critical to their cultural and political survival. However, they continue to lack both the financial and human resources necessary to run their own programmes successfully. It is essential that programmes be developed and adequately supported financially to build the competence of band decision makers, band staff and the membership. It is also critical that this be done in a way that effectively respects and links the traditional modes of interaction and decision making with the technical and analytical modes from the external system (Wolfe, 1989).

Fourth, the First Nations are in the process of becoming managers or joint managers of considerable lands and renewable resources across northern Canada. In claims settlement areas they have right of exclusive use and management of large tracts of land. They now have strong numerical membership on boards and commissions in the Territories, and there are proposals for Aboriginal participation in land and wildlife management in parts of the provinces. Some argue that this is transforming the north. However, the extent to which they will be able to command the research capability necessary to have reliable information on which to base their input into decision making is presently uncertain.

In many respects, emerging Aboriginal governments are tending to mirror the organizations that have administered them, and which they have to negotiate with to gain greater self-government. Aboriginal governments are being drawn into an increasingly rather than less bureaucratic system, with multiple relationships with other levels of government (including other Aboriginal governments), and with numerous committees, management boards and advisory boards, as well as policy and decision-making councils. Tensions develop between the bureaucratic structures and processes of contemporary government, and the consensus decision-making principles on which Aboriginal governments are traditionally based and the interpersonal relationships and informality of community tradition.

Although it is clear that change is underway in Indian and Inuit Territory, the prognosis for the emerging instruments of Aboriginal government is less certain. The new systems are burdensomely bureaucratic, and the bureaucracies that preceded them have not been dismantled. The governments have limited untied or independent sources of revenue through which they can exercise their own priorities. Although considerable, government funding to Aboriginal peoples continues to be inadequate to meet even basic needs. Aboriginal people are getting a stronger say in how lands and resources are used, but have outright ownership of a relatively small land base. There is little prospect of significant restructuring of land and resource ownership in Canada. Structural changes and a shift in power in favour of Aboriginal governments is a necessary condition for substantive improvements in Aboriginal communities: it is not, however, a sufficient condition. That requires a broad and consistent political will on the part of both non-Aboriginal and Aboriginal people and their governments, and commitment of substantial resources. Across Canada there is a groundswell of popular support for Aboriginal self-government: political will continues to be lacking.

References

Canadian Arctic Resources Committee (1988) *Aboriginal Self-government and Constitutional Reform: Setbacks, Opportunities and Arctic Experiences.* CARC, Ottawa.

Cassidy, F. (1991) First Nations can no longer be rebuffed. *Policy Options* 12, 3–6.

Cassidy, F. and Bish, R. (1989) *Indian Government: Its Meaning in Practice.* Oolichan Books and The Institute for Research on Public Policy, British Columbia.

Crowe, K.J. (1990) Claims on the land (part 1). *Arctic Circle* 1, 14–23.

DIAND (1982) *The Alternative of Indian Band Government Legislation.* Department of Indian Affairs and Northern Development, Ottawa.

Erasmus, G. (1988) *1987 Conference on Aboriginal Self-government and Constitutional Reform.* Canadian Arctic Resources Committee, Ottawa, p. 51.

Erasmus, G. (1989) Introduction and epilogue. In: Richardson, B. (ed.) *Drumbeat: Anger and Renewal in Indian Country.* Summerhill Press and Assembly of First Nations, Toronto, pp. 1–42 and 295–302.

INAC (1984) *Response of the Government to the Report of the Special Committee on Indian Self-government.* Indian and Northern Affairs Canada, Ottawa.

INAC (1986) *Alternative Funding Arrangements: A Guide.* Indian and Northern Affairs Canada, Ottawa.

INAC (1989) *Indian Self-Government Community Negotiations: Guidelines.* Indian and Northern Affairs Canada, Ottawa.

INAC (1990) *Self-government on Essential and Optional Subject Matters.* Policy Directorate, Self-government Sector, Indian and Northern Affairs Canada, Ottawa.

Jull, P. (1991) Canada's Northwest Territories: constitutional development and Aboriginal rights. In: Jull, P. and Roberts, S. (eds) *The Challenge of Northern Regions.* Australian National University North Australia Research Unit, Darwin, pp. 43–64.

Little Bear, L., Boldt, M. and Long, A. (eds) (1984) *Pathways to Self-determination: Canadian Indians and the Canadian State.* University of Toronto Press, Toronto.

Moss, W. (1985) The implementation of the James Bay and Northern Quebec Agreement. In: Morse, B. (ed.) *Aboriginal Peoples and the Law: Indian, Metis and Inuit Rights in Canada.* Carleton University Press, Ottawa, pp. 684–94.

Nishnawbe-Aski Nation (1986) *Memorandum of Understanding.* NAN, Sioux Lookout, Ontario.

Nishnawbe-Aski Nation (1991) *The Federation: M.O.U.* NAN, Sioux Lookout, Ontario.

Nunavut Constitutional Forum (1985) *Building Nunavut: Today and Tomorrow.* Nunavut Constitutional Forum, Ottawa.

Ontario Native Affairs Secretariat (1991) *Ontario Draft Paper: Issues Arising from the Nishnawbe-Aski Nation Model of Governance.* Ontario Native Affairs Secretariat, Toronto.

Penner, K. (1983) *Indian Self-government in Canada: Report of the Special All-party Committee* (Penner Report). Ministry of Supply and Services, Ottawa.

Weller, G.R. (1990) Devolution, regionalism and division of the Northwest Territories. In: Dacks, G. (ed.) *Devolution and Constitutional Development in the Canadian North.* Carleton University Press, Ottawa, pp. 317–34.

Wolfe, J. (1989) Approaches to planning in Native Canadian communities: a review and commentary on settlement problems and the effectiveness of planning practice. *Plan Canada* 29, 63–79.

Wolfe, J. (1991) Aboriginal self-government in Canada: current developments. In: Jull, P. and Roberts, S. (eds) *The Challenge of Northern Regions.* Australian National University North Australia Research Unit, Darwin, pp. 129–46.

CONCLUSION

Even the most cursory reading of the 20 chapters in this book reveals the complexity of rural systems and the significant scale of the contemporary changes taking place in them. Indeed a full appreciation of rural economy and society requires an understanding of a wide range of issues including: regional population age and size structures, national housing policies, changing life-style values, societal attitudes to the environment, competition for access to rural land, regional employment structures, and national and local development programmes for rural communities. It is not claimed that this book has investigated all of the relevant issues in the contemporary transition of rural systems. Rather, the assembled chapters highlight the need for specialist, but collective research effort by a team of investigators for an adequate examination of the complexities of contemporary rural economy and society.

To many of the authors contributing to this book, the changes, or transitions, found in today's rural systems are so significant that the term 'restructuring' can be aptly applied. While it will take the retrospective research of future analysts to provide an accurate perspective on current events, the last decade appears to have been an important period in the development of rural economy and society under advanced capitalism. For example, gone are the certainties that rural land must be protected for agricultural production, with implications for the employment structure of rural regions; new land uses have brought 'urban' people into the countryside, together with new attitudes and values on the rural environment; rural economies are no longer just linked to larger national economies – they now function in the context of an international or global economy; increasingly those who live in the countryside are interpreted as forming society 'in rural areas' rather than as a distinctive 'rural society'. This last aspect has theoretical implications for those who carry out their research in rural areas (see Chapters 1 and 2); but it is also reflected in the changed attitudes of national politicians and the policies they devise and implement for rural areas.

The origins of this book – in the papers of a conference – make it inevitable that some themes are repeated as various authors develop the logic of their arguments. However, insofar as the contributors have been drawn from three different cultural backgrounds (Canada, the United States and the United Kingdom), each with its own value system and assumptions, any repetition is itself of interest. It helps us to identify those commonalities that are found in the rural systems of all developed countries. In this respect, four main conclusions can be drawn.

First, as emphasized repeatedly in these studies from Canada, the US and the UK, the contemporary transition of rural areas has three principal dimensions, namely the transformation of the use of rural land, the social recomposition of the population, and the restructuring of rural labour markets. Taken together, these three dimensions both define the central structure of the 'rural system' and reveal the impacts of the processes of change; moreover, the dimensions are interrelated in the dynamics of the rural system. For example, changes in land use, such as the transference of agricultural land into conservation/environmental protection zones, impacts on the employment structure; the resulting movements of population seeking either new employment or access to 'desirable' rural environments lead to new social structures in the countryside.

Secondly, the (mega-trend) processes of transformation act unevenly on rural areas – by person, household, locality, region and country. Thus the contemporary transition of rural systems serves to perpetuate, if not enhance, variations in the quality of life between people and places. The chapters in this volume trace the source of this differentiation to two main, albeit interrelated, conditions: variations in cultural value-sets and institutional structures. In the first case, societies vary in the values attached to, for example, individual property rights, intervention by the national or local state, and the quality of the environment, including its aesthetic properties. In the second case, institutional structures over factors such as planning legislation, land zoning, agricultural policy, environmental law, housing policy and welfare arrangements for the disadvantaged, all act unevenly on rural economy and society – both within and between different countries.

Third, the 1980s have been a time when the perspective of 'bottom-up' community development has received greater emphasis relative to the previously dominant 'top-down' development paradigm. This is in keeping with the presently influential 'minimum intervention' view of the role of the state; but it also reflects the failure of previous 'top-down' programmes of rural development, together with a search for an alternative, and more successful, approach to the problem of uneven social and economic conditions. Placing the onus for development in the hands of the individual, or the individual community, divests national politicians and governments of that responsibility; but at the same time individuals are empowered to develop their societies according to their own value-sets. Of course this assumes a degree of unanimity within a community on the desired type of development, and a willingness to care for the disadvantaged in that community. The papers in this book are cautious on both assumptions.

Finally, and in keeping with one of the conclusions from Volume One, the theme of 'sustainable' rural systems recurs throughout the chapters in this book. The concept of sustainability can be variously applied to such issues as rural employment provision, the composition of rural society, levels of economic well-being and the structure of land-use systems. The extent to which capitalist market processes are able to deliver 'sustainable' rural systems is not confronted in these chapters, nor to any great extent elsewhere in the relevant literature; nevertheless, this important issue is worthy of further collaborative research effort.

Index